Mother Earth Is Our Elder

Books by Katłįą

Northern Wildflower

Land-Water-Sky / Ndè-Tı-Yat'a

This House Is Not a Home

Firekeeper

Mother Earth Is Our Elder

A Northern Indigenous Perspective on the Climate Crisis

KATŁĮĄ

McClelland & Stewart

Hardcover edition published 2026

The authorized representative in the EU for product safety and compliance is Penguin Random House Ireland, Morrison Chambers, 32 Nassau Street, Dublin D02 YH68, Ireland, https://eu-contact.penguin.ie

Library and Archives Canada Cataloguing in Publication
Title: Mother Earth is our Elder : a northern Indigenous perspective on the climate crisis / Katłı̨à.
Names: Lafferty, Catherine, 1982- author.
Identifiers: Canadiana (print) 20250315319 | Canadiana (ebook) 20250319365 | ISBN 9780771018572 (hardcover) | ISBN 9780771018589 (EPUB)
Subjects: LCSH: Elders (Indigenous leaders)—Northwest Territories. | LCSH: Traditional ecological knowledge—Northwest Territories. | LCSH: Climatic changes—Effect of human beings on—Northwest Territories. | LCSH: Climate change mitigation—Northwest Territories. | CSH: Dene—Ethnobiology—Northwest Territories | CSH: First Nations Elders—Northwest Territories.
Classification: LCC E99.A86 .L34 2026 | DDC 971.9/3004972—dc23

All photos courtesy of the author.

Book design by Andrew Roberts
Cover art: Beaver painting by Archie Beaulieu, courtesy of Theytus Books, Ltd.
Typeset in Radiata by Sean Tai and Six Red Marbles
Printed in Canada

McClelland & Stewart
A division of Penguin Random House Canada
320 Front Street West, Suite 1400
Toronto, Ontario, M5V 3B6, Canada
penguinrandomhouse.ca

1 2 3 4 5 30 29 28 27 26

Penguin Random House
McCLELLAND & STEWART

For my grandchildren, and all future generations

SeNǫ̀htsı̨, dìı dzę̀ę k’e gok’ènedì.

My Creator, today, look after us.

Hazhǫ̀ǫ ha Mahsì nets’ı̨ı̨whǫ,

We thank you for everything you give us,

Gozha, gokǫ̀, gonèk’e, hazhǫ̀ǫ ha

Our children, our homes, our land, everything

Wet’à ts’eènda.

With which we live.

Hotı hǫt’e nı̨dè.

Amen.

—WIÌLIÌ DEH YATIÌ K’ALEMÌ DENE SCHOOL PRAYER—

Contents

Mother Earth Is Our Elder

Languages

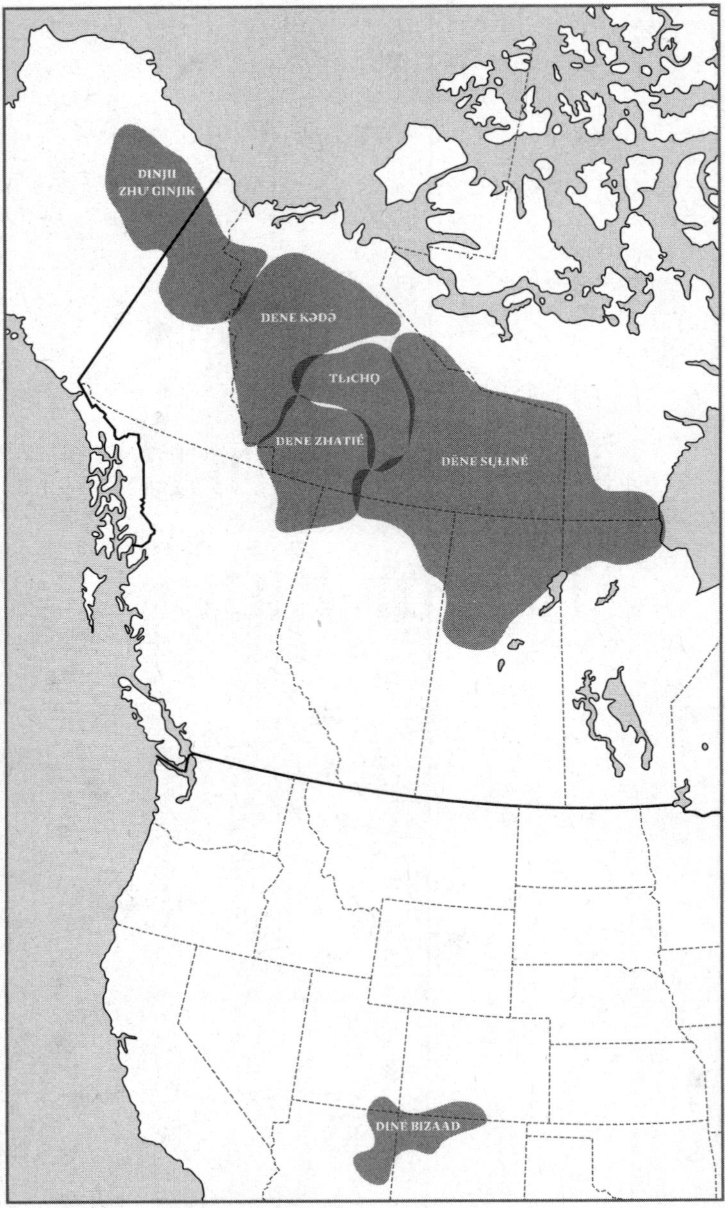

This map does not include all Dene across Canada.

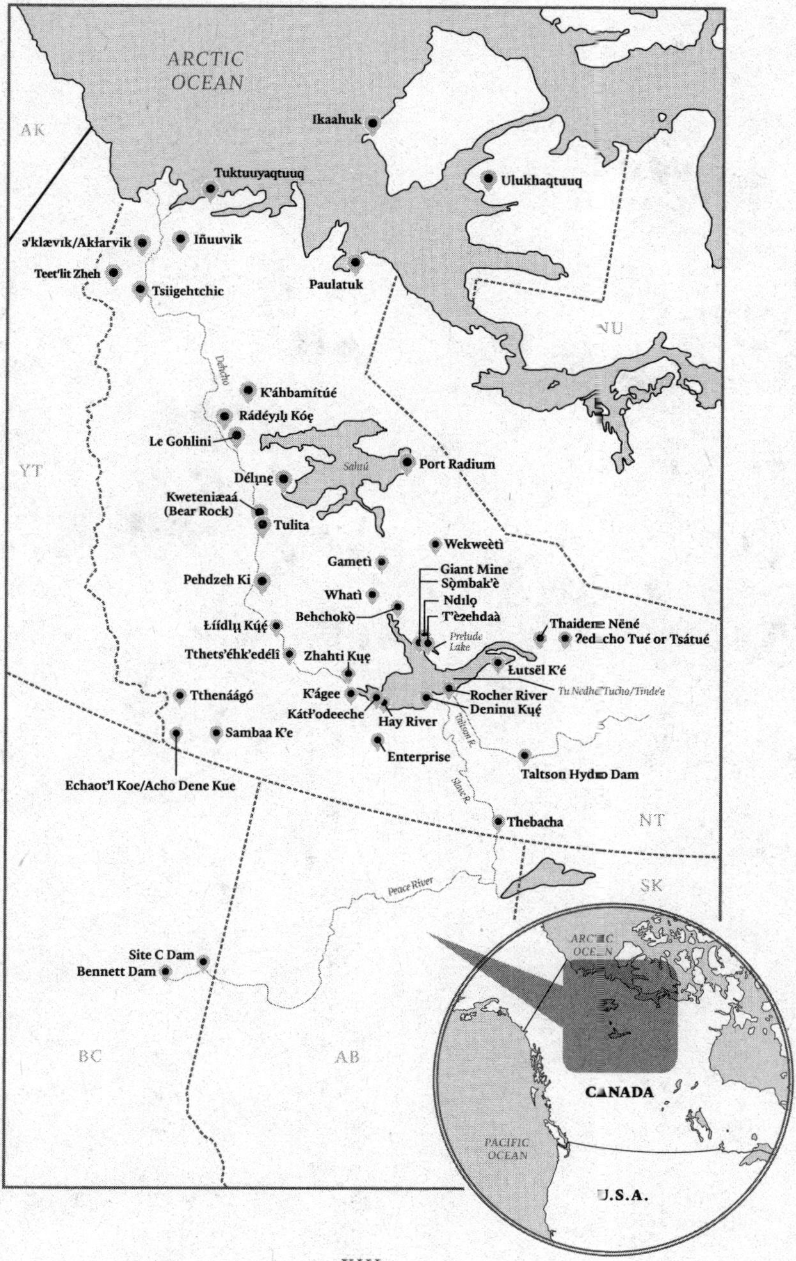

ARCTIC OCEAN
AK
Ikaahuk
Tuktuuyaqtuuq
Ulukhaqtuuq
ə'klævik/Akłarvik
Iñuuvik
Teet'lit Zheh
Tsiigehtchic
Paulatuk
NU
Dehcho
K'áhbamítúé
Rádéyı̨lı̨ Kóę
Le Gohlini
YT
Sahtú
Port Radium
Délı̨nę
Kweteniææá (Bear Rock)
Tulita
Wekweètì
Gametì
Giant Mine
Sǫ̀mbak'è
Pehdzeh Ki
Whatì
Ndılǫ
Behchokǫ̀
T'èɂehdaà
Łíídlı̨ Kų́ę́
Thaide Nëné
Prelude Lake
ʔed cho Tué or Tsátué
Tthets'éhk'edélî
Zhahti Kųę
Łutsël K'é
Tthenáágó
K'ágee
Rocher River
Tu Nedhe Tucho/Tinde'e
Kátł'odeeche
Deninu Kų́ę́
Hay River
Sambaa K'e
Taltson R.
Enterprise
Taltson Hyd Dam
Slave R.
Echaot'l Koe/Acho Dene Kue
Thebacha
NT
Peace River
SK
Site C Dam
Bennett Dam
BC
AB
OCEAN
CANADA
PACIFIC OCEAN
U.S.A.

Preface

The world is always changing. It is a gradual cyclical process of evolution over the passage of time. At one point, millions of years ago, the Arctic was a tropical oasis. Every continent on this planet we call earth was once connected. Mother Earth is in a constant state of movement, but it is changing faster than ever before. Humans have sped up this change significantly by not living in harmony with nature. An imbalance now exists in the ecosystem, and it is up to all of humanity to change the narrative—to help reverse the devastating effects this acceleration has had, or at the very least slow down the transformation for our own survival.

There are those who still don't believe in climate change or global warming, and there will always be naysayers. This book will not speak to you in numbers. I will not drone on with the scientifically proven data about climate change, but if humanity does not give up our reliance on fossil fuel, as of right now, the planet is projected to reach a 2-degree temperature change within the next decade and a 1.5-degree warming in the next few years. The only way to stop the planet from reaching this level of warming is if we shut down the production that is contributing to greenhouse gas emissions—yesterday. We had a window of opportunity with Covid, but we missed it. We need to slow down and Covid was a giant pause. We can expect that a 1.5-degree change will cause mass devastation that has not taken place in a very long time. With that projection to happen soon, a sense of impending doom exists for many, and some religious believers feel that it will be the second coming of Christ.

Whether or not we start to collectively make changes in our lifestyles to reduce carbon emissions, this shift has already begun. We may be able to stop the world from increasing further in temperature by making drastic, immediate strides to reduce our energy consumption, but it won't be enough. The world desperately needs a dose of strong medicine combined with the latest discoveries in science and advanced technology to provide a brighter future for generations to come.

In the Arctic, permafrost is melting rapidly, animal populations are depleting, water levels are fluctuating from flooding to drought, forest fires are out of control, and extreme changes in temperature make it unsafe for hunters to rely on seasonal weather patterns. My northern ancestors once set up camp as they followed the migration of the caribou, leaving a small environmental footprint, but this connection to the land has been severely interrupted. Indigenous people have lived alongside and protected nature since time immemorial. Through the generations we have had to stand strong and unwavering amid the unwelcome presence of industry that has resulted in environmental injustice and racial inequities. We have long protected Mother Earth from destruction through our intricate knowledge systems, natural laws and age-old principles. Now more than ever, the world is relying on the resilience of Indigenous people to help solve the climate crisis. It is my hope that this book will contribute to answering those calls for help.

Throughout this book you will see that I have used place-based names interchangeably between Dene languages and English. You will also see the words *human-made* change used interchangeably in place of *climate change* and *global warming*. The term human-made change puts the accountability back where it belongs: in the hands of those who have gotten us into this mess in the first place and those who have the power to stop our current ill-fated trajectory but choose to amp up production instead. The most powerful people in the world have the ability to stop global warming from advancing to the point of no return, but they are still operating as business as usual. If they don't stop production, we will all suffer.

Indigenous nations all over the world know what it takes to protect Mother Earth and its inhabitants for future generations. The answer may not be what you want to hear, but it's no secret: it's about letting go of unnecessary extravagances. If Indigenous peoples' knowledge of stewarding the land had been respected and heeded from the time of contact, we would not be in this mess. It is enshrined in most, if not all, Indigenous knowledge systems to not take more than what is needed, nor place emphasis on personal gain but rather uphold the value of nature over the accumulation of material wealth.

Today Indigenous knowledge needs to be not only shared orally but transmitted through the written word, because of the urgency of the climate crisis. Writing is a very different way of passing down our stories and is still frowned upon, but aside from the traditional methods of sharing oral histories, such as sitting around the fire, as we evolve we must adapt new ways of getting our message across to those who need it most, and incorporate the technologies that are made available to us to share our teachings for the greater good of humanity. We must mobilize Indigenous laws by sharing and living out our teachings everyday, getting us one step closer to a profound understanding of the purpose and meaning of life which is to live in a harmonious balance with ourselves, our families, our communities and the natural world.

You will come across passages in this book where I refer to a collection of written stories as told by Dene Elders living in the 1700s, called the *Book of Dene*. The stories were transcribed by a priest and there are instances where the priest interjected his own interpretations based on his religious beliefs, which has muddied the waters. It is important to keep in mind that the stories found in the *Book of Dene* are only a minuscule glimpse into the sheer magnitude of our storied history. I am merely a messenger gathering and sharing stories from my home far and wide. I believe I am very much taking after my many grandmothers. Especially my great-great-grandmother, for she too was a messenger. Her name was Catherine Beaulieu Bouvier Lamoureux. My English name is also Catherine. She is who I am named after. She was aptly known as Ehtsu Naats'I, meaning "grandmother of the winds," and she is deemed a historic person in the north. In the late 1800s she traveled across the north by dog team in the winters, delivering mail to local residents in neighboring communities and even bringing food to the children in residential schools. I believe that a part of her spirit lives on through me today, I may not have went by dog team but I too have traveled across the north to gather and deliver the messages found in this book. My Dene relatives kindly shared and entrusted me with their stories, their perspectives and teachings that were passed down to them by their Elders and they have done so because one of the teachings in our Dene Laws is to pass on the knowledge.

Half-way through the writing of this book I made plans to meet with Paul Mackenzie, a respected healer in T'èʔehdaà. We met at a local diner where the lingering smell of second-hand cigarette smoke hovered in the air from the rundown hotel upstairs. I used to have fries and gravy with my friends there as a teenager. I'd planned this dinner with my mom, my daughter and my Elderly friend Berna Martin, who was there to help translate for Paul, our guest of honor. It wasn't quite dinner time but it was already dark outside. That week, Yellowknife was experiencing a severe cold spell and I had to leave the truck running during the entire dinner for fear that it might not start without being plugged in. I felt like a hypocrite contributing to greenhouse gas emissions with an idling Ford F150.

When Paul arrived, we ordered our food and the first thing he asked me when I began the informal interview was "Why you?" What he was asking was why I was the one who had been tasked with collecting these stories and writing this book. It was a valid question. I am not a scientist or a researcher. I am a northern Dene woman who no longer resides in the north full-time even though I want nothing more than to live in a cabin in the woods on my home territory. I am far removed from being out on the land, unlike Paul and so many others I have spoken to in the making of this book, who were raised in the bush throughout most of their formative years and still spend every waking moment they can outside.

Paul's question brought me back to when I was a child and did not have a chance to spend time on the land learning about my culture because I grew up in an urban environment in low-cost housing and at times in the foster system. I fumbled with my words, trying to find an answer to his valid question. My teenage daughter looked at me provoking the question even further. "Yes, why you, Mom?" she asked, testing me. It felt as though everyone at the table was suddenly looking at me very intensely. I glanced over at Paul, who had a serious look. His face was bronze from being out in the sun all day, while I could feel mine turning a deep shade of red.

Why me? He could not know that the candor of his question cut through my thin layer of confidence and went straight to my self-doubt. I'd already asked this question myself countless times, but being asked by someone I respected and admired was different. Did he think I was the wrong person for the job? Who was I to write such a book? If this were the last book on earth, the one that would save humankind, then couldn't they find someone else for the job, someone with more experience? It was like he was looking right through me.

In that moment I felt as though the weight I was trying to carry by seeing this book through was too heavy and I was underqualified, but I was not going to let that stop me. Like my great-great-grandmother, who I'm sure battled more than a few storms in her life where at times she wanted to give up or did not think she had it in her, I will carry this bundle forward and deliver the message. Yes, it is a big responsibility, but it is my humble honor.

That is what I wanted to say but instead I told him the short answer, which was that I had written a proposal that was accepted by a publisher. I could have gone into a long and detailed explanation of how my work and educational experiences had led me to sit there with him at that table discussing the current state of the world, but it is not becoming to brag—our family is supposed to do that for us when we are not in the room.

I later learned that Paul had met with a group of Indigenous women from down south who call themselves the "Aunties." The Aunties had come up to Yellowknife, my hometown, for a visit and met Paul at a place called Dechinta—a center for research and learning in the north where I had worked for a time helping to lobby the government for funds for on-the-land programming. I happened to meet up with the Aunties one evening a few months after the dinner with Paul, and when they found out where I was from, they said that Paul mentioned how proud he is of me. That's when I realized, Paul may have only been testing me when he asked that pointed question "Why you?" Knowing that he is proud of me gave me solace and a boost of encouragement to see this work through.

This book could not have been written in a linear fashion because Indigenous stories are interwoven and complex. I found a way to weave each of the storytellers' teachings throughout, as the teachings cross over into many different aspects of life and cannot be categorized into chapters.

In the making of this book, I have sought out Elders,

community leaders, knowledge carriers and land guardians, like Chief Fred Sangris and Steven Nitah, who know so much about the places in which they are from. I spoke to youth like my son who is becoming an expert in forest fire management, and my cousin Randy Baillargeon, who is following in the footsteps of his grandfather Alfred Baillargeon who taught him nearly everything he knows about Dene ways of life. Their voices are integral to this book. Most people I spoke to I have known all my life, and some I was guided to—people like Mary Jane and Gilbert Cazon, who taught me so much in one single conversation, and Lawrence Casaway, who doesn't know just how much of a tremendous impact he has had on the direction of my life by telling me a story about a loon on a sunny afternoon. These conversations were deeply engaging and rich with knowledge. I will forever marvel at how captivating our stories are.

I spent time in many communities across the north to help inform the teachings in this book, from the far east of Great Slave Lake down the ramparts of the Dehcho by speed boat to Délı̨nę on Sahtú, and all the way out to the Arctic Ocean hitching a ride from Iñuuvik to Tuktuuyaqtuuq. I could not write about a place in depth without having been there myself and being immersed in the landscape, soaking in the natural beauty with my own eyes.

This book has been a healing journey for me that has only just begun. As much as I wanted to travel to every northern community, I was only able to travel to some—but I made efforts to represent as many Indigenous voices as possible. I did my best to meet with the storytellers in person, but there were times that I had to speak with them over the phone because I was articling for a law firm at the time and was unable to take time away from work to travel—which was beyond frustrating, because I wanted nothing more than to spend every waking hour visiting and talking to Dene storytellers and hanging onto their every word. Nothing else mattered to me. I spent most of my waking hours on the weekends and evenings putting this book together while simultaneously working in the area of solicitation, a field of law that was so dry I had to force myself to stay awake, and talked myself out of quitting several times—so needless to say I was grateful to have the much needed heartfelt conversations with the storytellers that made me feel that my life had purpose. You might notice that my relationships and connections are strongest close to where my family line is from, though I was able to travel as far as the Navajo Nation and even the Mayan ruins in Mexico to better understand our connection nation to nation, and what I have come to realize is that we have strong ties. I wish I'd had more time to devote to visiting and learning from community members, but these are unprecedented times and this book is urgently needed. If only the readers could benefit from hearing directly from each Elder and knowledge carrier who shared their teachings in this book while sitting with them and visiting, they would be much better for it. Yet in the midst of our busy lives we can always pause and reach for a book to provide us with the connection we seek and at the very least I hope that this book will offer a sense that you were right there with me on this journey.

Just as we are connected to one another, so too are we connected intrinsically to nature. We cannot exist without connection to one another and to the natural environment around us. In learning how a teepee is constructed I was told that the teepee exemplifies coexistence. A teepee cannot be built alone. The tree poles are lifted by many hands and they represent our community. The hides that form the cover of the teepee represent our immediate family, and the fire that burns in the middle to keep us warm represents our hearts that burn brightly. We can never let the fire go out or our spirit will grow cold, which is why we must remember to feed the fire often.

In these pages I also share some of my own personal reflections and experiences. I do not claim to have all the answers. In fact, writing this book has caused me to have more questions than answers. I ask that you try and suspend any disbelief and judgement you might be carrying with you before reading any further.

Before you set off on your journey north, I would like to broaden your understanding of the simple act of shaking a person's hand when parting ways. Since in my language we do not have a word for goodbye, shaking hands is a way of saying "safe travels" before setting out on a long trip. So, let us shake hands, figuratively, as you embark on this adventure.

Land Acknowledgement

I have become nomadic like my ancestors and travel back home to Dënéndeh—meaning "the land of the people" or "the land of Dene"—as often as I can. I consider the north a wise Elder that I visit and learn something new from each time I return. I often feel homesick for the place below the 60th parallel. Being home feeds my spirit, uplifts and re-energizes me. Home is medicine for my soul. Whenever I go home to Dënéndeh, I am given many gifts. Not in the sense of material possessions, but something much more valuable. Sometimes these gifts are so subtle they go unnoticed. I am gifted with enlightenment and a broader awareness of the natural world, but only if I am paying attention. Are you paying attention to what gifts Mother Earth has for you in the place where you live or the place you are from?

I wrote much of this book while residing on the lands of the Coast Salish in W̱SÁNEĆ territory. Here on the Pacific coast, I am surrounded by old-growth trees and the ocean breeze. I have lived on these lands for some time now and always make a point to learn about the history of this place through stories told by the first people. I honor the teachings and make it my responsibility to live as a respectful guest.

Just like I could not write about a place without ever having been there, I also could not tell these stories without being a Dene person myself, building lifelong relationships with the storytellers and having their permission to share their stories—which is what many non-Indigenous people fail to understand, particularly well-intentioned researchers. You cannot go into a community, extract knowledge and get the credit for it. Necessary measures must be taken, proper protocol must be abided by and permission must be first received. I went to great lengths to ensure that each storyteller in this book had a chance to review and approve their stories before going to print. I also made sure to compensate the storytellers in exchange.

There are certain unwritten protocols that are followed when one nation is crossing into another nation's territory in the north. It is not often that this permission is witnessed first-hand, but I had the honor of watching this consent process unfold before me a few years ago when serving as a council member for my First Nation, the Yellowknives Dene.

I was in a meeting in the community of T'èʔehdaà in the band council building that is shaped like a large teepee. It was a beautiful winter day. During the meeting, we all heard the sound of a skidoo engine out front. That is not an uncommon sound in the winter in the north, but then we heard another skidoo and another until the sound of skidoo engines were so loud we couldn't ignore it. There were ten of us on the council, including two Chiefs, and we all got up and walked over to the large windows that looked out over Tucho, the big lake, to see the spectacular sight of at least a hundred skidoos side by side waiting in a long line as if they were about to start a race.

There was some commotion with the administrators running up and down the stairs, I'm not sure if some phone calls were made, but what I do know is that it was incumbent upon the late Edward Sangris, Chief of T'èɂehdaà, to step outside onto the balcony alone. I thought he was simply going out to get a better look, but he was about to do something almost unheard of in this day and age—he was about to make a gesture that was once commonplace in our Dene communities.

The Chief went outside without a jacket in −30, and stood there while the drivers sat idling on their skidoos looking up at him from afar. He looked like royalty, like a king about to make a speech. Without a word, he waved and gave a nod. Just one nod and the skidoos took off to the east, to where they knew the caribou were. It was the Tłı̨chǫ Dene, neighbors to the Yellowknives Dene, and they had been respectfully asking the Chief for approval to cross through the Yellowknives' area—also known as Chief Drygeese territory—to hunt. That day I witnessed how the Tłı̨chǫ had not only asked for permission to cross the invisible boundary line across the land, they had also made an unspoken promise, without the need for a legal contract, to share with the Yellowknives if their communal hunt was successful. If this were to have taken place a hundred years ago, it would have been a line-up of howling dog teams in place of skidoos.

This is what true land acknowledgement can look like. Just as you would not go into another's backyard without permission as it is a trespass, you wouldn't go into someone's home without knocking first, somehow this protocol has been forgotten.

Growing up, my grandma taught me to always give an offering of thanks whenever stepping on new land. I am to put tobacco down on the ground or some other offering such as a coin, to ask Creator to keep me safe on my journey. I ask for each and every one of you to put an offering down on the ground wherever you are as we go on this journey together.

What would happen if everyone in the world paid their respects to Mother Earth in this way? What if Mother Earth had a giant guitar case opened in front of her for us to drop coins in as she sang her sorrowful songs on the empty, dirty streets of the city? What if we gave back to Mother Earth instead of constantly taking from her?

— 1 —

Laws, Language and Land

Every Indigenous nation in the world has their own laws, language and land. All have been interrupted in one way or another by the impacts of colonization, but all is not lost. Dene Laws, language and land are woven together like a thick braid.

The Dene Laws are a guide to live by. In the Northwest Territories (NWT), Dene ways of knowing and being are still strong today. The Dene Laws, in one form or another, are:

- Share what you have
- Help each other
- Love each other as much as possible
- Be respectful of Elders and everything around you
- Sleep at night and work during the day
- Be polite and don't argue with anyone
- Children should behave respectfully
- Pass on the teachings
- Be happy at all times

Many Dene collectively refer to their northern homeland as Dënéndeh. From the northern boreal terrain to the southern desert landscape, more than a dozen Dene groups exist across the continent. Throughout this book, however, I will refer specifically to the Dene of the NWT, a territory in the far northern reaches of so-called Canada where I am from. Dënéndeh, translates to "the people of the land" or "the land of Dene" in my language.

The word *Dene* means "flowing from Mother Earth," as *De* means "flow" and *Ne* means "Mother Earth." Dene also translates to the word "land" and/or "people" in most every Dene dialect across Dënéndeh. There are eleven official Indigenous languages in the NWT and five of those are Dene languages, which are:

- Dëne Sųłıné Yatıé or Chipewyan
- Gwich'in
- Sahtúǫt'įne Yatį́ or North Slavey
- Dene Zhatıé or South Slavey
- Tłįchǫ Yatį̀

As Indigenous languages continue to be revitalized, it is important not to be ashamed or embarrassed if the words are pronounced wrong—because that is what was done to the children in residential schools—children were taught to be ashamed of and beaten for speaking their language. It is vital that Indigenous people hang onto the languages that were once so cruelly ripped away.

That is why I proudly go by Katłįą now. Many people cannot pronounce my Dene name, but they try and that is what matters. I first began regularly asserting my Dene name when I started law school. As we went around the class introducing ourselves, it felt strange to say that my name was Katłı̨ą because I was so used to referring to myself as Catherine, but Katłı̨ą is the name that I have been called by my grandmother and great-aunts since as long as I can remember.

Mary Rose Sundberg, who goes by Maro, is a Wıìlıìdeh Yatıì translator from the community of T'èʔehdaà who founded the Goyatiko Language Society. She officially solidified my name one day as we sat around the table visiting after a language class that she was teaching. She wrote out my name with the proper syllabics on a napkin. Since then I have asserted my name in my language when introducing myself, as it provides a sense of identity and belonging that connects me to my home. I have thought about legally changing my name but I would rather not give my Dene name to the government.

The name Katłı̨ą is a mix between my English name and the Wıìlıìdeh Yatıì language of the Yellowknives Dene First Nation. The word Wıìlıìdeh means "fish river." Wıìlıìdeh can also mean "Yellowknife River." Yellowknives Dene otherwise known as the T'satsąot'ınę would gather year-round at the mouth of the river to celebrate, and still do to this day. The T'satsąot'ınę resided in the area long before their hunting, fishing and gathering grounds were encroached on by settlers. The T'satsąot'ınę translates to "metal or copper people," as we were once referred to as Copper Indians by those who made first contact with us because our ancestors wore copper daggers around their necks. Copper was very sacred to us. The words *Yellow Knife*, became one over time and the city of Yellowknife, where I grew up, is named after the Yellowknives Dene.

Whenever I introduce myself, along with identifying with my Dene name, I also tell people where I'm from, but first I tell them who raised me: my grandmother Alice Lafferty and grandfather Edward Lessard. I then tell them I'm from Sǫ̀mbak'è, which means "money town" in Tłı̨chǫ Yatıì. The town of Sǫ̀mbak'è or Yellowknife was built at the time of the gold-mining exploration boom, it was literally a money town. Now considered a city, it has grown to a population of approximately 25,000 people and is the capital of the NWT. Yellowknife is a transient place, mostly full of out-of-town government workers who have moved north to gain working experience that they cannot get elsewhere. But prior to the gold rush, the area that is now called Yellowknife was once completely inhabited by the Dene. Once settlers occupied the area, the Dene were no longer able to hunt moose or gather berries, even though what is now the center of the city was once a pristine blueberry hill that the T'satsąot'ınę harvested from annually. Now many of the Yellowknives live on the end of an island just down the hill from Yellowknife in a community called Ndılǫ. Like many Indigenous people across the continent, they have been pushed into a corner by settlers. Ndılǫ is where my mother was raised and still resides.

The Dene are a part of the larger Athabaskan-speaking people. The Athabaskan trail is said to have started 35,000 years ago, but it very well could have been much earlier—it's just that science has not found the evidence to back up our oral histories yet. We, the Dene, are holding strong at an estimated 45,000 altogether in this country we call Canada, and the

Dene in the NWT make up about a quarter of the population. Before contact, it is believed that there were more than ten million Indigenous people inhabiting the North American continent, with the Dene making up a large part of the population. Due to intertribal wars and violent conflict between the Dene and Europeans—who attempted to assimilate us into a set of different spiritual beliefs by way of unfair tactics, both sword and pen—as well as sickness and plagues, our numbers have dropped significantly over the passage of time.

Our relatives in the southern United States are the Diné. The Diné occupy modern-day New Mexico, Colorado, Utah and Arizona and are found in great numbers in urban areas outside of these states. The Navajo Nation Diné are the largest tribal legal system in the world, exceeding 300,000 citizens, and over half of them live on the Navajo Nation reservation. The Diné have principles that they live by which are embedded in their language just like the Dene. Both the Diné and the Dene consider spirituality, medicine and sacred ceremony as parts of their collective identity—and a large part of this is held in story.

Oral history forms the foundation of Indigenous laws. Indigenous stories are like an ecosystem—they are all connected like an intricate web. They flow into one another. What might look like a short story is often part of a much larger story. There are stories inside of stories in what some might refer to as a stream of consciousness. There is a story in every rock, tree, creek, river and lake. Stories ground us. Indigenous stories hold analogies of the natural unspoken laws of Mother Earth. Our stories exist inside of place-based teachings and are often told metaphorically. These ancient stories hold clues to an unimaginable past that is often far from what the average human mind is capable to comprehend because over time we have been taught to turn away from and downplay the importance of story, referring to it as fable. But the truth is that stories expose powerful messages and there are forces in this world that do not want us to have that power.

It is not uncommon to have similar teachings across Indigenous nations far and wide. Take our relatives in the south, for example. With the Dene being related to the Diné, whose name also translates to "people", we share very similar linguistics and commonalities among our stories of creation which can help us piece together our shared history. One of those common threads between the two groups is a story of twins, sometimes referred to as cousins, who were heroes for the people because they were able to kill giants. Across Dënéndeh these two hero twins are referred to as Yamǫ́rıa and Yamǫǫ̀zha. I have not been able to come across information about the mother or father of the two brothers in our Dene written or oral histories, but I was able to determine, in my limited research, that the Diné call the deities who bore them Asdzą́ą́ Nádleehé, meaning "Changing Woman," and Yoolgai Asdzáá, meaning "White Shell Woman." These deities may also be one and the same. The Shá, meaning "sun", had a part to play in bringing the twins to life. Together the two deities, helped to create the earth. Changing Woman is said to be the source and sustenance of all life, while White Shell Woman brought hope, protection, and a connection between humans and the divine. Could it be that Mother Earth was in fact first created by both man and woman and by non-gendered entities like the sun, and not only by man as we have been conditioned to believe by religious groups?

There are different stories about how the Dene and Diné separated long ago in the four directions of the earth. The *Book of Dene* tells of how division among nations came about and includes versions of the same story from the Dëne Sųłıné, the Tłı̨chǫ, the Gwich'in and the Sahtú. A story called "The Rainbow" tells of how, according to the Sahtú Region's K'asho Got'ine version—otherwise known as the "big willow people"—two brothers, most likely referring to our Dene heroes, were out walking together when a rainbow appeared. They ran toward it until they reached a mountainside, where there was an old man sitting in his hut. He asked what the brothers were doing, and they told him that they were running toward the rainbow, but in doing so they no longer recognized the landscape they were in. The old man warned them not to go near the rainbow. He then gave the brothers one arrow, and instructed them to go toward the setting sun and use the arrow to hunt but insisted that they not reclaim the arrow after it had been shot. The brothers went on their way and came upon a squirrel. The younger brother fired the arrow at the squirrel but did not listen to the old man's instructions. He went to retrieve the arrow once it was shot but the arrow moved on its own. The two brothers ran after the darting arrow until they were further and further away. They then arrived at an enormous mountain so large that it touched the sky. They followed the arrow to the summit in the sky. It was there that the brothers heard voices coming from inside the mountain, speaking about how their languages were different. There were many men assembled inside the mountaintop, and it was too small to accommodate the masses. A fire was lit and the rock began to split. The men were scared as the mountain crumbled and turned into a vast plain. They scattered in many directions until eventually new nations were formed. From then on, no common language was spoken among them.

My friend Steven Nitah is Dëne Sųłıné Yatıé and he knows where I can find more information about the great separation that occurred between the Dene and the Diné thousands of years ago. Steven is from a small community called Łutsël K'é. He was born and raised on the land with his grandparents and great-grandparents. They moved with the seasons and slept in a tent or a cabin. In the winter they set up camp in the boreal forest because it is a good place for trapping. In the spring he and his family spent their time along the shores of the Tu Nedhé, meaning "big lake" or "big water" in Dënesųłı̨né and otherwise known as Great Slave Lake. His family would harvest beaver and muskrat. He lived in the tundra in late summer and into the fall for caribou hunting. Steven didn't start school until he was ten years old. Up until being introduced to the English language through the school system, he only spoke Dënesųłı̨né. I thought I had many varied careers in my lifetime, but Steven has me beat. He has been a Chief, a band councilor, a member of the legislative assembly of the Government of the Northwest Territories (GNWT), worked in the mining industry, was an associate producer for a CBC television show telling stories across northern Canada in all different Indigenous languages, and in the last decade he's worked tirelessly for his nation and has been instrumental in helping to create a protected conservation area called Thaidene Nëné, meaning "land of the ancestors" in Dënesųłiné.

Steven, who is often pictured in the media wearing a flat cap hat and a moosehide vest with beaded flowers and feathers, explains that the word for Creator in the Dëne Sųłıné is Nǫ́łtsine. The word *nu* means "island" and *sine* is "earth created." Therefore, earth is an island created. If earth is an island, then the sea is space, and like any island—whether on a lake or ocean—there are finite resources. We have to be able to live in connection with those resources to maintain a balance, a relationship of reciprocity, says Steven.

The concept of earth being an island correlates with other Indigenous stories of when the world was covered in water. In some of these stories it is the muskrat who sacrifices his life, sometimes it is a duck or other water mammal that dives down to find soil, and brings it up to the surface of the water to form the continent that we now call North America which Indigenous people often refer to as Turtle Island because it is shaped like a turtle from above.

The Dene have a similar story that has been lost in translation over time but is captured in the story of the Dene Sųłıné called "The End of the World" in the *Book of Dene*. In this story there is a great flood that is caused by Squirrel, who releases sunlight which melts the ice age. In another version of the story, Squirrel saves the world from winter by releasing the sun after Bear puts the sunlight into a bag and hangs it in a tree on an island. The other animals wanted light, so they worked together to release the sunlight from the bag. Once the sunlight was released it melted the ice and caused a great flood. The animals that could swim dove down to try to find land. Muskrat tried, Otter tried, but it was long-tailed Duck who dove down and managed to come up with mud in its webbed foot and after that the water began to recede. The Sahtúǫt'įne Yatį́ version says it was Beaver that brought up the dirt. The Tłı̨chǫ version says it was Muskrat who came up to the surface with earth in his paw, in a story called "Tpkakfwele's Flood."

Keep in mind these stories are told by Elders who would have heard these stories as children by their Elders who were alive in the 1600s, which were passed down by their Elders who were alive in the 1500s and so on. What's not clear is how stories of the Dene and other Indigenous nations could have been so similar at a time when there were no cell phones, and no modes of communication other than contact through pre-colonial trade. Our stories would have traveled along our trade routes. Like a game of telephone, some things change over time but the origin of the story always stays the same.

The *Book of Dene* does not talk about how Diné and Dene separated but what I have been told is that a long time ago, when Diné and Dene people were once living together as one, they saw a vision of what was to come. Our people are known to be dreamers and visionaries. Some of us to this day are gifted with the ability to see visions of the future. Some would refer to this as a vision quest. At that time, long ago, the visionaries saw that strange-looking people with different beliefs were going to arrive and when they did, they would try to eradicate our race. We saw danger before it arrived.

The Creator gave us medicine that helps us to communicate in other realms, often through dreams. Our souls travel when we sleep. I myself sometimes have visions of the future in my dreams—mostly related to going to new places I've never been before, and I get the strong feeling of déjà vu once I arrive months or years after having the dream. I believe it's the universe telling me I'm on the right track in life.

When the news came that people were coming to disrupt their peace, four arrows were shot up in the air and we spread out in all four directions so that we wouldn't all be caught in the same area. Steven connects one of those symbolic arrows to a spot in Thaidene Nënë. It is a big rock that looks like it has a handle that is sticking up and out of the rock, near where giant beavers, or Tsàcho in Wıìlıìdeh Yatıì, were said to be located, at a place called ʔedacho Tué or Tsátué in Dënë Sųłıné Yatį—meaning "beaver lake," otherwise known as Artillery Lake. It is so named because there was a giant beaver living there. The location of the giant beaver lodge remains an important spiritual site for the Łutsël K'é, explains Steven. "When we travel on the lake, we stop there to make offerings."

It is important to note that there are many different understandings and perspectives on how our nations separated and more than one type of narrative should be considered. Nonetheless, Steven thinks that it is good that the Dene and Diné spread out, so that they can be a refuge for one another. He says that some Dene Elders have given instructions for young people to collect and store iron tools in deep water holes, because they will need to be relied upon if we have to go back to the stone age—if the world as we know it and all its technology shuts down. This is one of the reasons why the establishment of the protected conservation area, Thaidene Nënë, is important to Steven and his community, as there is a better chance of it being a refuge for future generations if the land is not spoiled by industry and development.

The Dene people in Steven's hometown of Łutsël K'é, which is a small fly-in community in the most north-eastern part of Tu Nedhé, are the guardians of Thaidene Nëné.

The first time I visited Łutsël K'é was in the winter in a small charter plane with a team of negotiators to provide the members of Łutsël K'é Dene First Nation—part of the Akaitcho Treaty 8 Tribal Corporation—with information on the progress made at the negotiations table with the federal government and the GNWT on the Akaitcho Agreement-in-Principle, a modern treaty agreement. When I spotted the community for the first time, I realized why it was once referred to as Snow Drift. The homes were situated at the bottom of a steep hill they looked like they might have just been freshly covered in an avalanche; the snow was piled high on each side of the community from the wind blowing in off the lake, you might miss it if you blinked.

The second time I went to Łutsël K'é, I flew with one of the council members from the Yellowknives Dene First Nation in the summer. We flew over a mountainous ridge called Betsų̨ghıé—meaning "the water is alive" in Dënë Sųłıné Yatı̨́. The large rectangular flat rock is also referred to as Utsingi Point, and is a chunk of land sticking out prominently from the middle of the lake that goes on for miles and drops off into a straight cliff on all sides.

I had the opportunity to visit the community of Łutsël K'é for a third time during the making of this book—shortly after I spoke to Steven. I just so happened to be asked by the GNWT to facilitate a community wellness event. At the Łutsël K'é airport I was greeted by the manager of the local Co-Op store; he drove me to a large blue house on top of a hill next to the police station. The local Dene call the police dëne nałtsí dëné, fittingly translating to "people taking people."

From every window in the blue house, I had a million-dollar view of the lake and a backyard view of a black three-legged dog sauntering by, sniffing at sticks. In the morning as I made my way down the hill to start my day a white bunny hopped along a trail and a bevy of snowbirds flew low and then high in a formation out of a nearby bush. I almost made it to the center of town but was stopped in my tracks by a flurry of fluffy husky puppies, who were huddled together under an old rundown pick-up truck. They ran up to me, their tails wagging as they attacked me with kisses. I was tempted to stuff one in my jacket and bring it home with me. I didn't care that they dirtied my jeans with their muddy paws. I always carry a dog treat or a piece of cheese when walking around the communities in case I run into strays. It's a fast way to make friends, especially when they are running after you barking wildly.

I found my way to the Co-Op store and stocked up on food for the few days that I would be there. In Łutsël K'é there are no Amazon or DoorDash deliveries but people do order fast food from Yellowknife and have it delivered on incoming flights. The small grocery store has nearly everything one could ask for—even fresh coconuts.

Most everyone in Łutsël K'é gets around on four-wheelers in the spring and summer and skidoos in the winter except visitors like me who have to walk everywhere. They cruise without helmets making sure to look out for the backwards water truck being pulled around by a tractor on its way to fill up people's empty water tanks.

The community wellness event was a success. Most of the town showed up for the feast and many stayed to discuss their concerns. When it was time to go home, I wasn't ready. I could have stayed longer but had to make my way to the airport to catch the plane back to Yellowknife.

I love Łutsël K'é and many of the other remote communities across the north because they carry a unique free-spirited vibe that is the opposite from cities down south where everything is so structured and orderly. Every once in a while I don't feel like picking up my dog's poop in a plastic bag and am tempted to just let it naturally disintegrate into the earth, and I'm tired of waiting at red lights, getting stuck in traffic and hearing sirens every few minutes. Traffic is like bad sinus congestion. I long to live in a place where I don't have to pay for parking, among the other expensive and unnecessary inconveniences that come with city life. Łutsël K'é has all that and more.

Łutsël K'é is one of the places across the north I would turn to for refuge if I knew the world was ending. I have been told that we will know the time is near because we are going to start seeing a lot of things in the sky. The sun and the moon will start to dance and the aurora borealis will begin to flicker in the day. These unusual events will play tricks on people's eyes.

The northern lights were in the south recently. They appeared one evening through a solar storm, an anomaly that seemed to have landed directly above my bedroom window. It gave me the comfort of home at a time when I was feeling homesick but it felt unsettling to see them dance above me in the sky in the south.

For me, the northern lights are our ancestors, they guided the Dene to the caribou herds. I like to think of the lights as a gift and a reminder to follow the light.

Will the earth give us more time? We are on borrowed time as it is. We are almost to the point of no return. The trash is piled up so high that you can't see through the garbage anymore.

The world is comprised of both natural and unnatural entities. All life on earth is made up of energy. Our energy vibrates out of our bodies, and if we isolate ourselves in a box the vibrating energy is kept inside the box. A house is just a box. Inside of that box our energy bounces off the walls and impacts other people. Energy is contagious. If someone inside the house is in a bad mood then usually everyone is in a bad mood. This is why it is important to spend time outside so any negative energy can be released into the atmosphere. "Touch grass" is the phrase often used by gamers who know this to be true.

There is a lot of energy floating around, especially in cities where people congregate. When people pass by you it is possible to pick up their energy, both good and bad, and that is why it is important to go out into nature to clear any negative energy fields from others that we may have unknowingly picked up—before they enter into our bodies, minds, spirit and home.

The English language that has impeded on our Dene languages can also be negative and condescending. Take the word *understand* for example, there are two words in one. Does it mean that you are supposed to be standing underneath something?

The English language was used as a weapon of deception—just look at the broken treaties that were signed all across

Turtle Island. Parliamentary laws are still written in old English purposely created to be hard to "understand" for most people.

The Europeans taught us and forced us to learn English words while our own language was left in the corner. Some have forgotten that we became humans because we could speak. Our language was ripped away from us to prevent us from knowing our identity and communicating with one another and passing down our stories, but we have still managed to hold on to our identity and our language.

Sometimes Dene language can be quite literal, and there may be words used to describe a word in English that would not make any sense without context. During that trip to Łutsël K'é I got to visit my Elderly friend Celine Marlowe at her home. She laughed as she told me about working as a translator while in college and coming across a place called Wild Bread Bay in English and the only way she could translate it into Dëne Sųłıné Yatıé was to say, "the bread is gone crazy."

I first met Celine Marlowe when I worked at Dechinta, a land-based program where the land is the classroom. We were at a staff retreat getting ready to bring the next cohort of students to camp, and she was hired to help tan caribou hides. I instantly admired her for her patience and her incredible ability to rapidly rappel down a steep cliff with only one hand on the rope as if she were riding a bull. Celine was brought up in Tacheè, otherwise known as Fort Reliance, which is a very sacred area for the Dëne Sųłıné people. She has since moved to Łutsël K'é as Tacheè is not a place where people live year-round anymore. Łutsël K'é is a hundred kilometers away from her birthplace.

When Celine was young her father sat her down and told her, "One day you're going to be teaching the language." She thought to herself, *Me? I'm only a kid.* All these years later his hope for her to keep the language alive has come true. Celine became a certified language teacher and taught for nearly forty years before retiring. During our visit she reminisced about her father as she listened to the small radio beside her on her kitchen table. A woman was speaking fast and fluently in Dëne Sųłıné Yatıé. Possibly one of her former students.

Another teaching that Celine's father instilled in her was to respect her Elders and to always share. "Even if it's the last piece that you have." For many Indigenous people, the more you give away, the richer you are. This is explained thoroughly in Potawatomi botanist and author Robin Wall Kimmerer's book *The Serviceberry*, in which she provides lessons on what is known as a Gift Economy. This too is one of the Dene Laws. Dene laws are not placed in any sequence of priority. All are equally important. However, one of the cardinal Dene Laws is to help one another.

There is a show called *Alone* that follows contestants who have to live alone in the wilderness, and whoever can survive the longest wins. One of its seasons was filmed just outside of Łutsël K'é. When the show begins, everyone is healthy and confident that they will be the last one standing, with many of the contestants bragging about how they are expert outdoorsmen. But it doesn't take long for people to start experiencing hardship and give up—before succumbing to a harsh environment where they are unable to catch enough to eat—and viewers watch as they begin to starve. The reason the contestants are unable to survive is the entire premise of the show: they are alone. They don't have community to help them, they don't have a support network around them. They have a man-against-nature mentality and they are not guided by the Dene people who have lived in that area for thousands of years and know how to survive.

The collective is important. Everyone is born with a gift to bring to the table. Not one person is good at everything. When the Dene lived solely off the land some were good hunters, others were good at catching fish, while others seemed to always find the best berry-picking spots and many still do. These different skills among the collective make for a diverse community to ensure survival. We need to be able to rely on one another. If the contestants on that reality show were able to incorporate the Dene Law of helping one another, instead of trying in vain to survive on their own in the wilderness without help, they would probably never have to give up and go home out of defeat. There would be one person designated to tend the fire, another going out to set traps or snares, another checking fish nets, another picking berries, another preparing meat and so on. This is how the Dene have always done things and how our ancestors lived in a harsh northern environment for thousands of years.

I grew up in a household where every one of my immediate relatives was impacted by intergenerational trauma, and even though I knew practically nothing about my European ancestry—I was still so colonized that I once brought a picnic basket on a boat trip and got made fun of by my family. It came equipped with plates, utensils, wine glasses and decorative napkins. Such luxuries are not found, nor appreciated, in the rugged wilderness, and wouldn't do me any good if it came down to a life-and-death situation. It's just extra weight. Needless to say, fancy decor out in the bush is—absolutely useless.

I don't know how to haul water, shoot a gun, or start a fire from scratch to keep me alive if I were ever stranded out in the cold. I am proud to say, however, that I can pluck a duck in record time. At one of our annual summer gatherings at the mouth of Wììlììdeh a few years ago, I entered the women's duck-plucking contest. Each contender had to run and grab one of the freshly shot black ducks lying in a heap at the side of a large round wooden platform which served as a stage where later that evening we would all dance around the central firepit in a drum ceremony. I ran as fast as I could, scooped up a warm duck, found a soft spot on the grass, knelt down and started plucking away at the feathers furiously yet delicately, being careful not to rip the skin. Plucking a duck is like plucking your eyebrows. The trick is to pluck in the direction the feathers grow not against them. The feathers were flying all around like I was in a mad pillow fight at a sleepover party and some of them even got into my mouth as I hurried to make it to the fire to singe the featherless bird. The race isn't over until the duck is plucked, singed to a crisp, cut up and cooked over the fire, ready to serve to an Elder.

I didn't end up winning the duck plucking contest, but I wasn't upset about it. I was honored to lose to an Elderly woman who had been plucking ducks a lot longer than I had. I remember when I was young my grandma would have random geese given to her and she would put them in the kitchen sink until they were thawed and ready to pluck. I wouldn't like to get too close or help her pluck them, and I certainly didn't care to eat duck soup either. But that was before I knew any better.

Deep within me lies the desire to learn how to make a life for myself out on the land, but I am one of the many Indigenous people who have not had the opportunity to be brought up in an environment where I can practice and live out my culture daily. This predicament of being dispossessed, where one longs for a home they've never known, is a lonely feeling. Even though I am part of a loving community, I still sometimes feel like I don't belong.

When I was young I only went to a drum dance once a year if I was lucky but I should have been attending all the drum dances, learning where the best berry-picking spots were and how to trap, hunt and fish. I should have been learning my language too, but my grandma was conditioned by the church, the royal family and settler society to believe that if she taught me these things she would be setting me up for failure in life. She was made to believe their way was better.

I could only admire the vast wilderness from afar when I was a child. This may be why I am sometimes afraid to be alone in the woods—because I am simply not used to it. We didn't have a skidoo, a boat, a cabin, or any of the modern-day necessities needed to enjoy the land when I was growing up. I was lucky if I had a brand-new pair of boots. These luxuries were always out of my family's reach. Money was hard to come by for us because we didn't place importance on it.

Nowadays, those able to experience the great outdoors have the means to afford it. It costs money to rent a campsite, to put gas in the tank, and to fill up the cooler. People pull up to campsites in their recreational vehicles and plug in, and some of them have television screens bigger than the window they should be looking out of. It seems ludicrous to concede that one must be able to afford the ability to "rough it," but somehow roughing it has turned into *glamping*.

It never used to cost anything to go out on the land. In the book *We Remember the Coming of the White Man*, a compilation of stories told by Elders from the NWT about what it was like living in the far north in their younger days, one Elder recalls when money first came to his community. To him, money looked like buttons. It was useless out on the land.

It took hard work to live on the land. Not many people these days know how to skillfully build a canoe out of birchbark—peeling the soft pliable birchbark, rolling it into bundles, soaking it in water, weighing it down with heavy rocks until it dried flat, and collecting spruce roots and splitting them with a knife before using the roots as thread for sewing the canoe together but my ancestors did. I rely on this detailed explanation as told by the late Elder George Blondin in his book *When the World Was New* because I have not seen it done myself and it's not something that's searchable online with step by step how to instructions—it's something that's taught.

Most people in my community no longer have the time or the know-how to build a birchbark canoe, construct a cabin out of logs and moss, or skin, stretch and tan the hides of a caribou, because colonial systems attempted to erase those teachings. While many of us have become accustomed to urban life, having to work a mundane nine-to-five job to make a living, there is still a large majority of Dene that live with one foot inside and one foot outside on the land, and subsist on locally harvested food. Fresh-caught fish and caribou meat is still hung to dry over smoldering fires in our communities on the shores of two of the biggest lakes in the world. Dene languages are still fluently spoken, and legends are still very much alive under the aurora-filled night in winter or the midnight sun in summer, prophesizing the precipice of change.

I reached out to Gladys Norwegian, former Grand Chief of the Dehcho Region of the NWT, when I was working on a story about northern energy and what she said has always stayed with me when I think about what the word wealth really means. Wealth, she says, has nothing to do with material gain. The health of the land is wealth. If there's one thing that impoverishes people, it is having the spirit snuffed out of ourselves." She says. One of the things she learned from her brothers is that "the land is still strong for us, but people have lost the way. Us as human beings have lost our way. We must somehow find a way to reconnect."

The health of the land is wealth.

If it weren't for the Dene acting as guardians and making it their sole priority to ensure that the NWT remains pristine, the majority of the land would be ravished by those looking to turn a profit and overlooking the north's extraordinary wealth. Since first contact, we have been removed from our culture through residential schools under the umbrella of the larger colonial regime, and made to believe that land does not hold meaning so that we don't feel a sense of loss when it is destroyed for monetary gain. Indigenous peoples especially have been intentionally forced to disconnect from nature, because once we are disconnected we are easier to manipulate—making it easier to take over and control the land and the people on it.

It has become our duty as Indigenous people to lead the way, and our laws and language are our weapons of resistance to government and industry incentives in exchange for access to our lands, that is why land, language and law are intertwined.

Indigenous leaders are under pressure to give up their land rights in exchange for financial incentives like unequal impact benefit agreements. We are given jobs when new mines are built but these jobs are often only entry level and seasonal at best, leaving no opportunity for job security. This is perpetuating the government-created system of dependence and control that stems from the monarchy and prevents our communities from becoming sovereign again—instead leaving us no choice but to rely on government contributions as a form of control.

With revolutionary movements like Idle No More—where Indigenous nations made significant headway when uniting as one voice to assert sovereignty and the need for environmental protection, holding protests that originally started in Canada and spread like wildfire throughout the world started by Indigenous matriarchs—Jessica Gordon, Sylvia McAdam, Sheelah McLean and Nina Wilson—it might seem like our voices have finally been heard, but oppression is still alive it's just less obvious than before.

We must continue to assert our Indigenous laws, language and land so that we can rebuild the broken parts of ourselves and our communities that have been chiseled away by capitalism over the last few centuries. We know that change is not going to happen overnight. It's taken so long to get into this mess that it might take just as long to get out. It's exhausting to have to constantly fight for what's right but we have no other choice and we are not giving up. A resurgence of Indigenous law through language and story, will arm us with a defense for the protection of our lands from further destruction in the name of capitalism, and through it all we will form unbreakable bonds with one another along the way.

— 2 —

Our Home On Native Land

There are three distinct northern territories in Canada. The NWT is not to be confused with the other two territories, the Yukon and Nunavut, which is usually the case whenever I tell people in the south where I'm from and they often say, "oh I love the Yukon." The NWT used to be the largest territory in the country because it was combined with Nunavut. Now it is the second largest territory in Canada and the second largest land mass over all provinces.

The NWT is home to two of the world's biggest and deepest lakes, which are arguably connected through groundwater reservoirs. The NWT is just over 1 million square kilometers in size with a population of a little over 40,000 people. This leaves a lot of wide-open space along the vast terrain. The NWT comprises five regions and twenty-seven Dene First Nations. It is no place for the weak. For centuries, survival here has depended on knowing how to live solely off the land.

The five regions of the NWT are:

- The North Slave Region, the ancestral homelands of the Tłįchǫ and the Dëne Sųłıné

- The South Slave Region, where many Dehcho people and Dëne Sųłıné reside
- The Dehcho Region (meaning "big river" in Dene Zhatıé) or Mackenzie River
- The Iñuuvik Region or Beaufort Delta Region, also known as the Gwich'in Region
- The Sahtú Region (*sah* meaning "bear" and *tu* meaning "lake" in Dene Kədə́)

You might be thinking to yourself, why do two of the regions in the NWT contain the word *slave*? There are mixed understandings of where the term *Slavey* derives from, but the most rational explanation I have come across in my research is that it was introduced by French explorers—as the word *Athabaskan* translates to *esclave* in the French language, which could have over time evolved into Slave or Slavey. This theory makes sense as there were relations between the Dene and the French Europeans, including my grandma's father who was from France. Another explanation is that at one point in history the Cree and Dene were at war and the Cree enslaved some of the Dene people. For what it's worth, there have been calls to change the derogatory wording but because NWT politics are so entrenched in bureaucratic red tape the excuse is that it is too hard to make the change in the span of one term in the legislature.

Not only does the name Slave or Slavey need to be changed back to its original name, so do many places across the north including the names of streets that were named after European explorers. Work is being done to reclaim our place names starting with reverting back to the original place names of our communities. In total, there are thirty-three communities across the NWT. Many NWT communities are remote, meaning they are only fly-in or accessible by winter road when the lake freezes over, allowing vehicles to drive across. Some government officials are of the view that the people living in these remote communities should just amalgamate to the capital of Yellowknife.

But traditionally the Dene never lived in one large shared space. Instead, we were very spread out and lived within our immediate family lines only coming together as nations once or twice a year for celebrations or to have important meetings. There are many villages where families once lived across the NWT that are not listed as official communities by the GNWT—such as my grandma's birthplace, Nı̨hshìı̨, meaning "big mountain island" and otherwise known as Old Fort Rae.

The following is a list of all official NWT communities, which many of the storytellers in this book call home:

- Behchokǫ̀ ("big knife" in Tłı̨chǫ Yatıì) or Fort Rae
- Délı̨nę ("where the waters flow" in Dene Kədə́)
- Deninu Kų́ę́ ("moose island" in Dëne Sųłiné) or Fort Resolution
- Echaot'l Koe or Acho Dene Kue ("people from the land of the giants" in Dene Zhatıé) or Fort Liard
- ə'klævık/Akłarvik ("barren ground grizzly place" in Inuvialuktun) or Aklavik
- Enterprise
- Gametì ("rabbit net water or lake" in Tłı̨chǫ Yatıì)
- Ikaahuk ("place where one crosses" in Inuvialuktun) or Sachs Harbour

- Iñuuvik ("place of man" in Inuvialuktun)
- Rádéyį́lį Kóę́ ("place of rapids" in Dene Kədə́) or Fort Good Hope
- K'ágee ("between the willows" in Dene Zhatıé) or Kakisa
- K'áhbamítúé ("ptarmigan net place" in Dene Kədə́) or Colville Lake
- Kátł'odeeche ("willow grass river" or "hay river" in Dene Zhatıé)
- Le Gohlini ("where the oil is" in Dene Kədə́) or Norman Wells
- Łíídlįį Kų́ę́ ("the place where the rivers come together" in Dene Zhatıé) or Fort Simpson
- Łutsël K'é ("place of the cisco fish" in Denésoliné Yatį́/Dënesųłįné Yatį́/Dënë Sųłınë́ Yatį́/Chipewyan) or Snowdrift
- Ndılǫ ("end of the island" in Wıìlıìdeh Yatıì)
- Paulatuk ("place of coal" or "soot of coal" in Inuvialuktun)
- Pehdzeh Ki ("clay place" in Dene Zhatıé) or Wrigley
- Sambaa K'e ("place of trout" in Dene Zhatıé) or Trout Lake
- Teet'lit Zheh ("place of the headwaters" in Dinju Zhuh K'yuu) or Fort McPherson
- T'èɂehdaà ("burnt point" or "ash point" in Wıìlıìdeh Yatıì) or Dettah
- Thebacha ("beside the rapids" in Denésoliné Yatį́) or Fort Smith
- Tsiigehtchic ("at the mouth of the iron river" in Dinju Zhuh K'yuu)

- Tthenáágó (“strong rock” in Dene Zhatıé) or Nahanni Butte
- Tthets’éhk’edélî (“water flowing over rocks” in Dene Zhatıé) or Jean Marie River
- Tuktuuyaqtuuq (“resembling a caribou” in Inuvialuktun)
- Tulita (“where the waters meet” in Sahtúot’įnę Yatį́)
- Ulukhaqtuuq (“place where one finds material to make ulus” in Inuktitut) or Holman Island
- Wekweètì (“rock lake” in Tłįchǫ Yatıì)
- Whatì (“marten lake” in Tłįchǫ Yatıì)
- Sǫ̀mbak’è (“money place” or “money town” in Tłįchǫ Yatıì) or Yellowknife
- Zhahti Kųę (“the mission house” in Dene Zhatıé) or Fort Providence

As you can see, Indigenous place names are much more interesting than English. Indigenous place names hold descriptive meaning.

My grandma always told me that everything has a spirit. Every place is special in its own right, every place has a history, and every place has ancestors watching over it.

Elder and family friend George Mandeville knows this to be true as well. When visiting with him over coffee in the early afternoon on a sunny summer day at a place called Prelude Lake, George told me the story that his late friend Alfred Lockhart from Łutsël K’é shared with him. Keeping true to Alfred's version, George says that Alfred and his family were traveling by boat from Yellowknife to Łutsël K'é and pulled into a place called Narrow Island to wait for a storm to pass. Alfred decided to go for a little walk alone after a scrumptious dinner of fresh caught fish. As he was walking, he spotted someone coming down the hill. The figure was about a hundred feet away. Alfred and the man both looked at each other but Alfred felt that something was off. "This guy was not dressed like we are today. He was in caribou skins. Must have been from one hundred years ago."

Alfred got close enough to the man that he could have conversed with him, and in hindsight he said he was sorry that he didn't, but as the figure neared he realized that the man was somebody from a past life. "It looked like he was going to the water to get something." They stopped and looked at each other, and before the man turned and walked back up the hill and faded away—"just disappeared," says George.

George asked Alfred if he was scared and he said no. He said it's the type of thing that happens all the time when out on the land. It turns out that Narrow Island was historically a place where people went if they were stuck in bad weather, some may never have made it home.

Even the city has a spirit, and sometimes that spirit can be sick. When working in the inner city, I saw the drug epidemic up close. My office at the law firm was located in the heart of downtown in one of the biggest cities in Canada. I was not raised in a big city, so I was not complacent to what I saw. The first time I walked the streets of the city, I was shocked. There were people slumped over garbage cans as people in business suits walked past them pretending not to notice. The entire perimeter of the building I worked in reeked of urine, because local shops wouldn't let the homeless in to use the washroom. Shop owners sprayed a water hose to chase drug users away in an effort to keep the streets clean. One man let his dog relieve himself next to a man who was questionably dead lying in the middle of the sidewalk.

My boss wanted me to have my professional headshot taken one day for the law firm's website. I met with the photographer outside the office in the middle of the afternoon. At the same time there was a man a few meters away from us, clearly in distress. He had his entire life strewn across the sidewalk spilling onto the road covered with human feces and bodily fluids. He was pacing back and forth until the police came and detained him. They forced him onto the ground and handcuffed him. All the while I was told to smile for the camera as though nothing out of the ordinary was happening behind me.

There is no truth in the city. For centuries, cities have crumbled. There are some exceptions. There are ancient cities that managed to be sustainable for long periods of time, but all cities eventually meet demise in one way or another. They have always destroyed themselves or been destroyed. Even though humans tend to naturally congregate, the city separates us. The only truth that exists is in nature. Indigenous peoples have been called "savages," "uncivilized" and "primitive," but the only place I see these words come alive is in the city. Australian Aboriginal scholar and author Tyson Yunkaporta, in his book *Sand Talk,* says "the exponential destruction caused by cities feeds the exponential growth of infrastructure and population... in order for economic growth to occur, there must be more demand than supply." To break it down, there must be more people needing basic goods and services than there are goods and services to meet their needs. Put another way, there are a lot of people missing out on goods and services in the city. "There is no equilibrium to be found here," writes Tyson. There is an imbalance in the city that can never be corrected because a city is constantly needing to feed itself more and more. A city is not progressive because a city has no destination or improvement other than to temporarily gentrify and repeatedly regurgitate cosmetically.

Cities are like a conceited person who never wants to age—they need constant upkeep, require trending makeovers to keep from becoming decrepit. Nature is more and more beautiful when left alone. Mother Earth ages gracefully. She does not need constant maintenance to be beautiful. And just because the city is a concrete jungle doesn't mean the land underneath it is dead; nature finds a way in between the cracks in the concrete, vines crawl up abandoned buildings, rats and mice scurry across restaurant countertops at night looking for crumbs. Nature always finds a way and we have seen this with Covid with how quickly nature took over. Creatures and living organisms from the ocean washed up on shore and no humans were there to clear the beaches free of debris; vehicles were off the roads, allowing animals to come out of hiding; vines sprawled across walls, taking back the spaces they had been choked out of. Mother Earth has shown us that she will continue on, and in fact flourish by being able to replenish herself without our destructive human presence.

When all of the office buildings were shut down it made us realize that we now have the technology to work from anywhere and don't require office space. Perhaps all those empty office buildings can now be opened up to house the unhoused.

The city takes. Nature gives. There are too many people living in densely congested urban spaces. If people lived more spread apart and consumed locally there might be enough to go around. For instance, if more people opted to eat naturally grown food from the surrounding area that they lived in instead of resorting to fast food, which is condoning the unethical mass-murdering of animals, then fast-food chains would eventually become obsolete because corporations rely on supply and demand. We have the power to boycott; we just aren't doing it, because a quick meal is much easier and cheaper than going out and harvesting our own food.

Some cities are planning for more green space. Instead of parking lots we are seeing rooftop gardens. It is finally being realized that nature is vital to humanity. It's a simple fact that trees in cities provide canopies of shade and homes for birds, and they clean and cool the air. There was a time when houses were built from the very trees that were on the homeowner's lot. Roots and branches next to houses have become a nuisance to homeowners, because they can clog pipes and gutters so instead of living with nature it is destroyed for our own comfort.

Thankfully, outside of the city there are still entire families of trees keeping us alive. These trees are all connected. There is an intricate root system underground, and science is finally catching up with Indigenous knowledge systems that have described the communication that occurs between trees.

I was once told by a medicine healer that you can learn all you need to know from a tree by watching what goes to it and what flows from it. The simple act of observing how an army of ants march over and under roots working steadily, how a squirrel takes the seeds from a pinecone and plants them in the ground so new trees can grow, or the way a caterpillar chews a leaf, the way a bee takes pollen from a flower at the base of a tree, or the way a bird lovingly builds a nest for its young is life unfolding before our eyes. I invite you to find a tree and sit and watch it for an entire day to see how alive a tree truly is. It is mesmerizingly meditative.

In the north, the wind carries the ancient songs of my ancestors through the trees and settles on every branch and fallen leaf. My ancestors knew the land like the back of their hands. I am merely one generation away from being born in a teepee, but I spent much of my childhood living in a cramped one-bedroom apartment with my grandma, sister and aunt, oblivious to a life lived out on the land.

I now have the honor of traveling across the north later in life, and each place I journey to is special in its own right.

When out on the land a few summers ago on a small island called Enodah—meaning "many lynx"—I walked along the shoreline with an Elder who pointed out significant historical land-use areas. She showed me where a circular formation of rocks were situated that once held the hides of a teepee in place so that it wouldn't flap in the wind. As we walked further along, she pointed to a row of rocks that would once have been placed in the bottom of a birchbark canoe to flatten it.

These subtle markings would have blended into the scenery had she not shown me that they were the evidence of our ancestors who had lived there long before us. Next time you're out on a simple hike along a trail just know that you might be walking over ancient artifacts or cultural sites without even knowing it.

The places we are from have significance. The places we are from can ground us in who we are and give us a sense of identity. The land is our relative.

When our original place names were replaced, we lost a part of that connection to the land—and that is why it is so important to repatriate not only our artifacts but also our place names, so we can once again find our way back to ourselves. The trees, rivers and lakes are our family. I have come to understand this from Dene Elders like husband and wife Gilbert and Mary Jane Cazon from the Dehcho. They have been together so long they finish one another's sentences. There is an entity in everything, they say. Even in rainbows, says Gilbert. "Once you see this, you never have to feel alone," Mary Jane adds.

The land is happiest when there are humans around, says Tłı̨chǫ knowledge carrier Maurice Zoe. "Nowadays there's nobody in the bush. The trees are crying." People used to set rabbit snares on the trail and talk and laugh and make the land happy. But now there are no people climbing hills and admiring the view. "The spirit in the bush is crying for people."

I believe that Nı̨hshìì is one of those places that Maurice speaks of. Nı̨hshìì misses that connection to family. Nı̨hshìì, my grandmother's birthplace, was one of the first communities in the north where the Hudson Bay trading post set up shop. It was once a bustling village with fur traders coming from all over the world but now no one lives there anymore. It is a beautiful piece of land that was formed in the Paleozoic era, hundreds of millions years ago. It is in the shape of a perfect circle from above because it is an old volcanic crater made of Acasta Gneiss, the oldest rock in the world. To know that my grandma was born a twin on a volcanic crater in a teepee makes me proud.

It is at my grandmother's birthplace that I spent many summers as a teenager reuniting with relatives during our gatherings. One year, I was invited to assist in conducting an archaeological dig to provide evidence that the Dene were the first inhabitants of the area and to uncover answers about my family's history. We carefully set gridlines and spent a few weeks digging up the remains of an old house, where we found flint, arrows, a stone hearth, and many other artifacts to prove that people had lived in the area for centuries.

The fields of archaeology and anthropology, and basically any profession ending in "ology," for that matter have historically operated more like an exclusive members-only club to which Indigenous peoples do not have an invitation. These white-centric elitist fields of academia denied Indigenous people the right to keep their own artifacts, which in turn also denied us the right to be able to tell our own stories—while those who polished and placed the bones of our ancestors in museums to be curated reaped the benefits of telling a one-sided clandestine history. Some historians have even gone as far as insinuating that Indigenous peoples were completely wiped off the face of the planet and no longer exist so they could fit their narrative into a larger agenda that gives governments an excuse to take over Indigenous lands.

Referring to our stories as myth rather than reality keeps the truth hidden, because if the truth is exposed it has the power to change the course of history and disrupt everything that we've been taught as a collective society by colonizers. Indigenous stories hold truth, but in order to have our oral histories taken seriously by the powers that be, we still need the archaeological evidence to back them up.

To date, 6,500 archaeological sites exist in the NWT. Very few archaeologists, paleontologists or anthropologists make an effort to work with Indigenous people of the north to unearth clues of their past. This is very problematic because many archaeological sites could be lost forever once exposed to the elements under the conditions of rapidly melting permafrost. After listening and learning about the stories of her ancestors, my daughter now wants to be an anthropologist—and I'm proud of her for taking an interest in her culture.

It's important that my daughter has people to look up to in this field of work. People like Letitia Pokiak, who is one of the few Inuit female archaeologists in the world. Letitia's traditional name is Panikpak and she is from Tuktuuyaqtuuq (Tuktoyaktuk), a small community along the shores of the Arctic Ocean within the Inuvialuit Settlement Region. Letitia was raised in her culture and spent a lot of time as a child on the land learning what it means to be Inuvialuit, living sustainably and in harmony with the land and animals. In her formative years she attended Grollier Hall, one of the last residential schools open in Canada. At the time it was called a "hostel." After high school Letitia got a Bachelor of Arts in

Anthropology, and after having children she went back to school and got her Master of Arts in the same field. I first came across her work when looking for Indigenous northern environmental activists to sit on a climate-change panel discussion. Her research at the time focused on the coastal erosion in her home community.

Letitia has been involved in a number of archaeological projects within the Inuvialuit Settlement Region. She was involved in an expedition on Banks Island that located the HMS *Investigator* wreckage, the ship captained by Robert McClure that had been sent out in search of Sir John Franklin's lost expedition. She also excavated an igluryuaq, or sod house, at an ancient Inuvialuit village called Kuukpak located along the Mackenzie River. She brings an Indigenous perspective to these sites, to prevent harm that could be caused by development or human-made change. This is not just a job for Letitia, it is her life's work.

There is a site where the Mackenzie River meets the Arctic Ocean which has been impacted by erosion and exposure to natural elements over time. It was there, Letitia informs, that a piece of pottery was found. This is a significant find because pottery artifacts in the north are few and far between, and bringing it to a lab to study it further is what would have occurred in this case, but Letitia was advised by her Elders to leave it where she found it, to not dig up the past, and so that is what Letitia and her team did. They left it on the ground.

Letitia's Elders believe that artifacts and ceremonial sites such as burials are a part of the earth and are to be left alone letting nature run its course. She has been informed that artifacts are not to be tampered with because they are someone's belongings and should not be rummaged through by strangers.

This viewpoint turns the whole concept of archaeology on its head. Before speaking with Letitia and learning from her, even I was of the view that if human-made change threatens to erode these precious artifacts then we need to act fast, but when put into the context of what her Elders are saying it's less about removal and more about respect.

Chief Fred Sangris of the Yellowknives Dene First Nation for the community of Ndılǫ would agree. Fred is the son of a man named Tachį̀į̀, meaning "drifted snow." Fred's father was born on a windy day in the spring of 1927 at Tacheè, their family village. Fred was named after his father—his Dene name is Tachį̀į̀ Wezhaà, meaning "son of drifted snow," because he was the firstborn son in his family. When he was baptized, Fred was given his English name.

The beginning of Fred's life was nomadic, cultural, and full of teachings from history and legends that go back hundreds of thousands of years. Fred was born into a family of storytellers and has learned many lessons from many Elders throughout his life, and so he has a lifetime of knowledge that I am only barely catching up on. Each time I speak to him, I learn something new.

I first met Fred when we worked together on the Akaitcho agreement. One day in my office he told me all about a historically documented massacre at Bloody Falls along the Coppermine River between the Inuit and the Dene—and how the story, as told in Samuel Hearne's journals, was one-sided. He pointed at a map on the wall showing where the massacre had taken place and spoke so clearly about the event as though it happened just yesterday, as though he had been there himself and lived to tell the tale. It was Hearne and his men who had kidnapped a young woman which triggered the massacre between the two nations.

There is a message that Fred wants to share about how wrong it is to take from the land. To illustrate this he tells me a story about how, in the 1970s, a woman found a copper dagger at the very end of Ndılǫ. The woman kept the knife until the day she died. When her children were going through her estate, they turned the knife into the museum in Yellowknife, where it is now on display.

Fred has many other examples of this "finders keepers" mentality that he says many settlers have. When Fred found out that random people were going to the shores of Wììlììdeh with their metal detectors and taking artifacts they had dug up from the ground, he told the museum it needed to stop. He tells me there is also a man who lives in Yellowknife who has in his possession the skull of a Dene person with a bullet hole. He was told that the man had even flashed the skull around at one point, showing it off. When Fred heard about this, once again he marched back to the museum and told the custodians to do something about it, but he says as far as he knows no one has investigated. This man is not the only one who is holding hostage the bones of our ancestors. Other people have been seen using Indigenous skulls as ashtrays, Fred tells me. This is an abuse of the remains of our ancestors and is a crime.

People find things all the time in the north, but many people "have a habit of taking things," says Fred. When people find something that does not belong to them, they should turn it in respectfully, he says. Thankfully, not everyone is dishonest. Fred knows of one man who was paddling the East Arm of the Great Slave Lake from Tacheè to Sǫ̀mbak'è. About halfway, the man got caught in a storm and had to pull ashore. He was walking on the side of a hill and happened upon a birchbark basket. Inside there appeared to be a knife made out of bone that was about a foot and a half long, lying next to a gill net made out of red willow. "It was there for a very long, long time," says Fred. It was very brittle. The man carefully brought the basket and its contents back to Sǫ̀mbak'è and turned it in. Now it's at the museum on display. "It's over two hundred years old. This guy did the right thing." Now the descendants of whoever made that net can view it, but "if it got into the wrong hands it would have been destroyed."

One time Fred took his Kwet'įį̀, "white" friend, on the land to go moose hunting. They were standing on a hill looking over the landscape when Fred asked his friend what he saw, testing him. His friend looked around and said, "I see money." Fred laughed, but his friend continued excitedly, "There might be gold in that hill. Look at those big trees, lots of timber."

Right there and then Fred knew they had different ways of looking at the world, and it is because of this type of mindset that the earth is getting abused. When Fred says that he and his friend "don't speak the same language," he is not only referring to language in the literal sense, but also to not having the same basic values when it comes to caring for Mother Earth.

There are places across Dënéndeh that are treeless, places that are lush, places that are rocky, and places that have a desert-like quality while other parts of the landscape look tropical with white sand beaches minus the palm trees.

Dënéndeh is treacherous and uninviting to those who don't know their footing. I have heard stories of underground rivers in Nı˛hshìì where fish without eyes exist, so when Chief Fred told me about sinkholes in the area I was intrigued.

In the north arm of Tu Nedhé, otherwise known as Tucho or Tinde'e for Great Slave Lake, it is all granite rock country, while on the southwest side of the lake it is limestone country. "Same lake, different environments," Fred says. The south side has underground streams and in some places there are sinkholes. Fred's father and grandfather used to tell him that if he ever goes hunting in that area, he must be sure to carry a long rope and not walk close to other people but to keep a distance because you can fall through the ground. "In the wintertime you can see steam coming up from underground but in the summer you can't see it." Says Fred.

Fred recalls traveling to the south side of the lake with one of his Elders from Ndılǫ named Jonas Noel—to a place called Ehdaalà in Wıìlıìdeh Yatıì, which means "White Beach Point." Jonas said to Fred, "Follow me. I want to show you something." The two men walked until they came upon an underground stream. The water was very cold. In that place there is no water as far as the eye can see but Fred says that when you get to the top of the hill there are little holes here and there that you have to watch out for because that is where the openings are to the groundwater. Fred believes that this underground stream goes on for miles and miles.

There is one particular location near the turn off to Whatì on the highway going south that has a natural hole in the rock which Fred says is from when the Tłı̨chǫ people first arrived on earth. They walked through that rock and said, "We are in a new world now." That hole in the rock is still there. "It's like a cave." Not far from it is a pit full of snakes, says Fred. "Most people who don't travel the land wouldn't know these stories, quite possibly at their own peril."

I am one of those people. I ventured out to find this hole in the rock on more than one occasion and still have yet to find it, it has evaded me. The first time I went was in the winter. I was with my daughter-in-law who was nine months pregnant and the snow was too deep. The second time I went was in the summer. I was able to drive the narrow dirt road leading to the edge of a cliff overlooking the tree line. I could only drive so far until the road stopped so I got out and walked the rest of the way. When I got to the edge of the cliff, the earth beneath my feet did not feel safe. I could see the remnants of where part of the cliffside had crumbled. It looked like an angry giant had come and smashed the landscape right where I stood. There was a deep gaping hole in front of me that looked like something could have been living in it. Suffice to say I didn't go on any further. Next time I try to find the rock I will ask someone who knows the area well, someone like Maurice Zoe or his brother John B. Zoe to guide me to it. On my way back down the gravel road I noticed how small the trees were. They were shorter than me. They were new trees and they were so healthy and so dense that I could hardly see through them. A forest fire had swept through the area and the trees were the new growth.

Forest fires have always swept through our territory but they are worse than ever. In 2023, a 417,000-hectare fire burned down the town of Enterprise before heading straight toward the city of Yellowknife, causing a mass evacuation resulting in the deforestation of 3.4 million hectares. When it was safe to return after the fires, mushroom pickers from the south came in droves. Ashes that are left behind after a fire enrich the land, and eventually there is an abundance of new growth—and in that new growth there are endless fields of morel mushrooms worth a lot of money. Fred, being the Chief, became concerned and felt it was his duty to warn the mushroom pickers that they should carry rope with them when out on the land, and not to walk side by side but to follow in each other's footprints just like his father had taught him. There was one woman in particular who was studying a new type of mushroom and was curious about it. She went off the highway about fifty kilometers past Edzo, a small Tłı̨chǫ community adjacent to Behchokǫ̀. Not heeding Fred's advice, she walked off the trail on her own and fell in a hole. "The hole was not that deep. She was able to get out of there. She crawled out." This happened after Fred had warned her, and his laugh when he tells me about it sounds like an "I told you so."

The woman came and visited him after she fell in the hole and said, "Fred, you're right. I fell through that hole. I was by myself but thank God I was able to get out." She then asked him if there are a lot of places like that and he said yes. Even in the olden days, when families used to travel or when they were coming back from hunting near Nı˛hshìı, they would tie a scarf or a belt made out of moose or caribou called a "whe" around their waists. "They would hang onto each other and that's how they would find their way back late at night. There were no flashlights in the olden days Fred says with a laugh." It's a very different country. "It's limestone, underground water, caves, everything you can think of. It's very, very different. It's almost like a different world when you go there."

As Fred is telling me all this over the phone, I can hear small birds chirping in the background. He must be sitting outside on his balcony enjoying the last days of summer, I think. That's when Fred switches gears and mentions we are not the only ones in that part of the country. "There's other creatures that live there." He explains that those other creatures have supernatural abilities. An Elder, who Fred calls an "old man," once told him that a long time ago those creatures had a leader, and he led them to the north arm of Tu Nedhé every summer. One time they took a person, and as they were traveling back into the mountains the leader said they were going to cross the river and that each of them was going to have to step in the same spot on the surface of the water or they would fall in. "They walked across the water," says Fred. "That's how they travel across river and lakes." No one knows where they are located but there are mountains to the west. "Big mountains, and the old man says that's where they come from and when they take you, you never come back." This is similar to the stories my grandma told me when she wanted to make sure I didn't run around at night when visiting Nı̨hshìì.

Fred's description of the northern landscape is on point. There is a very different topography in the north There are also salt plains and quicksand. As we sit across from one another out in Prelude Lake, my family friend Elder George Mandeville tells me about the time he was in the La Loche River area when he was a teenager. Being in George's company, you are guaranteed to hear many good stories. He was born in Fort Resolution in the 1940s, one of seventeen siblings, and is the last living son.

As I listen to him tell me stories, I put the coffee percolator on the gas stove. Prelude Lake is about thirty kilometers outside of Yellowknife, and the house I'm staying at runs off of an intermittent mix of solar and a generator for power. George only takes milk, no sugar. I'm not used to using an old tin percolator, and I made the coffee too weak on the first go around and had to add more and put it back on the stove for a bit longer. George likes his coffee strong.

"I was helping to pack up some gear that was left behind when I was asked to go get some water at a little lake close by," George recalls. He did as he was asked and ran over the hill. There was no one else with him when he jumped off a six-foot drop onto what he thought was solid ground and sank deep below the surface but still didn't hit the bottom. It was quicksand.

"I was up to my shoulders in this quicksand and I just turned around and there was some willow that just happened to be hanging, growing in the crevice of the rock." There were a couple of Elders about a quarter of a mile away, but George contests that they wouldn't have heard him if he yelled for help. "I was just able to reach the willow and grab on. I was careful not to rip it out but it was almost out of reach. Had I been another few inches past I would have sunk right out of sight."

When George made it back to camp he told everyone what had happened and they were just as surprised as he was. "You only see that in movies." He laughs. "It was like a porridge." No one had ever warned George about quicksand. "I was lucky." "God only knows how deep it was, because I didn't hit anything solid at all," he recalls. "I was totally in shock."

In the north you often hear stories of how people go missing when out on the land, maybe that's one of the ways they disappear without a trace or maybe it's one of the creatures that Chief Fred describes.

Quicksand, sinkholes and giants that can walk on water seem to be the stuff of movies, but Chief Fred says long ago there were giant beavers in Wıìlıìdeh. They came up from ʔedacho Tué. "Back then the people lived in fear, there was no peace. They didn't know when something big was going to walk into their camp or tip over their canoes."

That's when the two brothers, Yamǫǫ̀zha and Yamǫ́rıa, were called upon to help the Dene, says Fred. Yamǫ́rıa lived in the Sahtú and toward the Hudson Bay, while Yamǫǫ̀zha lived in the Yukon and Alaska, near the mountains and ocean.

Yamǫǫ̀zha was called upon by the Dene people at a time when large creatures walked and flew around the earth, and he was able to use medicine power to turn himself into a giant and walk across rivers and lakes just like the creatures. Once, Yamǫǫ̀zha broke up a dam that the Tsàcho—the giant beaver—built, and chased three beavers from Artillery Lake all the way to the mouth of the Yellowknife River. The beaver made a new dam and the water started to rise and rise, and so Yamǫǫ̀zha was called again. He pounded on the beaver house with a snow shovel to get them out, and the beavers swam out so fast that they pushed aside their dam and it turned to rock, and that is why today there is a point near the Wıìlıìdeh site that marks where the dam once was.

The beavers swam into Tucho and ended up at Nı˛hshìì. Yamǫǫ̀zha went after them again and chased them out of the area. The three beavers started to build a dam at the mouth of the Dehcho otherwise known as the Mackenzie River, Canada's largest and longest river, and the water went higher and higher. It was there that Yamǫǫ̀zha turned himself into a giant and kicked the beaver dam into the Dehcho and it turned into islands. The beavers swam down the Dehcho with Yamǫǫ̀zha hot on their tails until they reached a place near Fort Norman where the Dehcho and the Sahtúdə́, meaning "Bear River" in Dene Kədə́, meet. There Yamǫǫ̀zha shot one arrow at the beavers but missed. "The arrow was standing in the water for a very long, long time, many generations of people have seen it," says Fred.

Yamǫǫ̀zha then chased the beavers to Norman Wells and killed them. "He skinned them and cooked the three beavers. He put the beaver hide on the side of a mountain by Tulita because that part of the hill faces the sun where the sun is the hottest." When he was cooking the beavers, the oil from their meat went into the earth—and now, says Fred, all that oil has been taken out. Le Gohlini, Norman Wells, is the oil capital of the NWT. Before oil was struck in Alberta's tar sands, Imperial Oil was drilling it out of the ground in Le Gohlini in the 1920s. It is no coincidence that this is the exact location where the legend has it that the oil from the beavers went into the ground.

The well-known Dene legend of Yamǫ̀ǫ̀zha and Yamǫ́rıa and their mission to eradicate giant animals has been scoffed at by scholars. Even I was skeptical when I first heard the story. I believed the Tsàcho to be nothing more than just a mystical legend. Sometimes it takes seeing to believe, and that is exactly what happened when researchers found a *Castoroide* in the Yukon Territory. If you don't know what a *Castoroide* is, you might want to sit down. It is a giant beaver, roughly the size of a full-grown black bear and it roamed the North American continent up to 1.4 million years ago. Further evidence indicating that we have been here for millennia. Is it any wonder that the beaver is Canada's national symbol?

Chief Fred says that after the two brothers set out to heed the call of duty at a young age, time passed and they didn't see one another for many, many years. It was only when they were very old and had long white hair that they met again. The two brothers had a fight near Nıdítagh Túé—in Dënë Sų̨łıné Yatį meaning "barren shoreline," and otherwise known as MacKay Lake—out in the tundra toward the ocean. They wanted to know who had the most powerful medicine. They fought for days and days, but at the end there was no winner; they were both equal. Neither of them had more power than the other. That was the last time they spoke to each other. Yamǫ́rıa went west, says Fred. He ended up across the ocean somewhere near China, and turned into a big island to avoid his brother, that is what the old people say explains Fred.

Meanwhile, when Yamǫ̀ǫ̀zha was very old he was being chased by what is known as the "Evil One," who was a bad medicine man. The last story we hear of Yamǫ̀ǫ̀zha is that he went to Tuktuuyaqtuuq and walked across the ocean to Sachs Harbour. Yamǫ̀ǫzha's wife was with him.

Fred emphasizes the fact that Yamǫ̀ǫzha's wife had on a shawl. They too were going to an island like Yamǫ́rıa, but they were fleeing the Evil One. Yamǫ̀ǫzha felt that he could not get away from the Evil One on earth and so he had to go up into outer space, where he could be there forever and people could see him and believe that he used to be on earth. The Evil One followed, but he couldn't go up into the stars so he shot his arrow at Yamǫ̀ǫzha and pierced him close to his kidney. "If you look at the stars today you can see Yamǫ̀ǫzha's constellation. He's bent over. He has his arm in the air where he got shot. The arrow is still there, you can see it on his back. Right behind him walking up there with a big shawl is his wife accompanying her husband. They are there forever. The Evil One couldn't get there." Yamǫ̀ǫzha left unmistakable evidence for humans to see and know he was once on earth and to remember that his purpose was to help people, says Fred. It is said that one day Yamǫ̀ǫzha will return in human form.

Yamǫ̀ǫzha and Yamǫ́rıa appear in many transformation stories among the Dene in the north. Even though there are many versions, each story leaves behind clues in the remnants of prominent landmarks—like the sacred tree that once stood on the shoreline of Wıìlıìdeh that is considered to have once been used as Yamǫ̀ǫzha's shovel. The tree is said to have stood for thousands of years across from the Wıìlıìdeh gathering site. That is, until lightning struck the tree and fell around the time of the Covid lockdown. It was a critical moment in history for the Yellowknives Dene, for it has been said by our knowledge carriers that when the tree falls, it will mark a significant shift in the world as we know it. The tree now lies flat in its resting place, at the Wıìlıìdeh gathering site where people go to pay their respects and place an offering.

> "If you look at the stars today you can see Yamǫǫ̀zha's constellation."

On the day the sacred tree fell, Chief Fred was looking out his back door when he got a phone call from someone telling him the news about the tree falling. Fred went to the end of Ndılǫ with his binoculars. He saw two lightning strikes hit the water 100 feet apart. He had never seen that before at that time of year. Not long after that, he got another phone call saying that a baby had been born. It was then that he knew that something special had taken place.

Stacey Sundberg is the mother of the child who was born on the morning that the sacred tree fell. She is also the daughter of Maro, the woman who wrote my Dene name on a napkin. As the storm raged outside the hospital window, Stacey's labor contractions struck at the very same time as the lightning.

Stacey takes a moment out of her busy morning to speak to me over the phone.

She says she told her ehtsı̨, grandmother, Mary Louise Drygeese that she wanted to name her baby Kǫ̀naıtł'ıì, meaning "lightning" in Wıìlıìdeh Yatıì, and she agreed that name suited him best. Dene people are often named based on their natural surroundings and the elements that present themselves when a child is born. Stacey says that the naming of her great grandfather, Itò, on her maternal side, had a lot to do with naming her son in the Dene language. "It's so important that we revitalize Dene names and Dene ways of life for our children." She says.

Kǫ̀naıtł'ıì—Kǫ̀ for short—is still a young child but Stacey says, "He's a hunter, he's a helper, he's everything." When speaking to her, they had just returned from Łutsël K'é, where he set snares and fleshed hides with his grandma on his father's side, Celine Marlowe, the rock-climbing Elder from Łutsël K'é. "Everything we did—he did," Stacey tells me proudly. "He even hauled wood." When I ask if she thinks he might be a great leader one day, she says yes—if he is built up for it.

He's a special boy, agrees Chief Fred. Something remarkable happened that day. Yamǫ̀ǫ̀zha might be involved, muses Fred. "We will see."

— 3 —

Life-Givers

Before visiting Iñuuvik, the Beaufort Delta was the only region in the NWT that I hadn't been to. Iñuuvik has one streetlight in the center of town that swings haphazardly on a low-hanging power line. No one buckles up when driving around, and after a while the annoying seatbelt warning that goes off every few seconds becomes background noise. Shortly after landing in Iñuuvik, I was offered a ride from a woman who saw me standing alone outside the airport looking confused. "Oh no no," I replied, "that's okay but thank you." I would have taken her up on her offer but I knew it would have been out of her way, as my friend lived further out of town. But like the woman who offered me a ride at the airport, I soon learned that most people in Iñuuvik are very friendly and inviting.

I had traveled to Iñuuvik in hopes of meeting with a few Elders and knowledge carriers, but I didn't have any solid plans. I reached out to a few friends who lived there to see if they would show me around town, and that's when I was invited to a special event, a women's sewing circle that just so happened to be taking place in the community that week. It felt serendipitous that I had been invited, even if it was at the very last minute, and I was pleased that I'd left my schedule open to be able to attend. The only real plan I had was to spend time writing on my friend's porch and possibly visit some of the surrounding communities if I could pay someone to drive me to Tsiigehtchic, Teet'lit Zheh, and possibly take me down the newly constructed road from Iñuuvik to Tuktuuyaqtuuq—the first all-weather road to the Arctic Ocean. It was hard to find someone who wasn't busy to take the time out of their day to travel a few hundred kilometers, so when my friend told me the sewing circle was taking place and that I could attend as an observer, I put writing and road-tripping on hold and gladly accepted the invite.

The building the sewing circle was taking place in was called Nihtat, meaning "among each other" in the Gwich'in language. When I arrived at the sewing circle there was a large pile of shoes in the entryway of the building. Everyone takes their shoes off when going inside office buildings in Iñuuvik; it encourages moccasin-wearing, said one of the women when I commented on the mound of shoes in the lobby.

There were about thirty women in the room—a mix of young and old, some were with their grandmothers and I noticed the woman who offered me a ride at the airport was there too. The women were from all over the NWT, and a few were even from the Yukon and Alaska. I arrived a bit late. The door was closed, and I slowly opened it to peek my head inside. The women were all sitting in a circle, they saw me and gestured for me to come in by pulling up a chair for me so that I could join the circle. They were in the middle of going around the room and sharing stories. When they were done sharing, the facilitator asked me to introduce myself. That's when my imposter syndrome kicked in. I only knew how to sew at an intermediate level. But after I told them who my grandmother was and how she'd taught a women's sewing circle in Yellowknife every Tuesday for fifteen years at a place called the Tree of Peace, I felt more at ease because many of the women knew her. She'd even taught some of them.

As a child I used to go with my grandma to most of her sewing classes, but instead of sewing I would sit under the table and draw pictures of houses, and now here I was all these years later once again sitting in a sewing circle and not sewing. History repeating itself. My grandma never forced me into doing anything I didn't want to do, and I must not have shown an interest in sewing because I don't remember ever having a needle in my hand.

One of the women in the circle shared openly about how my grandma was the best sewing teacher in the north, and how she remembered a time when there were two white women in her class trying to learn to sew and one of them had accidentally poked my grandma in the arm with a needle when she was reaching out to pull her thread. The woman was mortified and repeatedly apologized, but my grandma just laughed it off. Just like my grandma's sewing class, the day spent with the women in that room was filled with laughter and helpful, caring hands. I felt like my grandma was right there with me in that circle. I was where I needed to be.

As they went around sharing stories another woman told everyone about how she took extra care of all of her supplies, even the scraps that fell on the floor. She said if she dropped her scissors she would find herself apologizing to the scissors.

Everyone laughed when she told them this, but there was something important behind her sentiment that was not lost on the women in that room—they all knew how important it is to care for everything we touch, because there is power and intention behind each and every one of our actions. All of the women in that room, young and old, knew deep down, as if it was second nature, that it is important to be respectful in all that we do but especially when creating something.

Our ancestors were frugal. They would spend many days and nights turning sinew straight from the caribou's flesh into thread that would be used to sew clothes for survival. This is far removed from today's fashion industry where everything is tossed in the garbage without a thought, where clothing made out of synthetic materials is made by the hands of children in forced labor. The materials used in fast fashion don't break down, which is bad for human health and bad for the environment. Whereas in my grandma's younger days, every part of the animal was used—right down to the gristle. "The new sinew is not the same as it used to be," many of the women remarked. They talked about how they did not like working with the new artificial sinew, but knew it was still important to acknowledge every piece of thread and scrap of material—no matter how big or small, new or old, real or fake. Seeing that level of care for their workspace made me think of my daughter and how she cares for inanimate objects as though they have a spirit. She would tell me that she felt sorry for rocks, or anything that she bumped into. She apologized to the dust when she swept it up off the floor and into the garbage. She knew then, even at such a young age, that everything has a spirit. The scientific name for this empathic nature is "anthropomorphism." Everything has matter and everything matters.

The women in the group mentioned the importance of ensuring excellence in their finished products and taking pride in their work. Some of them would spend a long time sewing together a piece of hide only to have to take the stitches out again because they didn't feel their stitching was good enough. It remind me of a conversation I'd had with Elder Mary Jane Cazon from the Dehcho. She told me, "It's all basically just teaching you about patience."

Mary Jane was not one of the women in the sewing circle, but she knows all about seeking near perfection when it comes to sewing. "Sometimes we rush through things but there is no need to rush," she says. If you make mistakes in your sewing, it is said that you will not have a good home—you are rushing through things. It was as though the women in the sewing circle fundamentally knew this to be true. My grandmother must have known it too, because I remember she would always tell people to start over, "undo the stitches and do it again" she'd say. Some of her students would be frustrated with having to take apart hours of their work, but now I understand that it was for their own good. Stitch by stitch, women weave their lives together. Instead of reading tea leaves or going to a psychic or fortune teller, Dene women can create their futures through sewing.

In the sewing circle I got to observe the different skills that each woman carried. Some were gifted at beading, while others were better at cutting out patterns and piecing them together. They shared their skills and relied on each other for their different strengths, helping one another. At the closing of the sewing circle, all the women who had completed a pair of mitts put them in the middle of the room to showcase them. Each pair was beautiful in its own right.

If this is what happens when a small group of Indigenous women get together, imagine what would happen if every Indigenous woman in the north gathered together in one big sewing circle? All of their loved ones would be warm and cared for at the very least.

When my daughter once asked me if she could have her own pair of moccasins, I felt a small triumph. The shame I once carried ended with her. When I was her age, I felt embarrassed to wear the moccasins and parkas my grandma made for me because I got bullied for it at school. Being included in the sewing circle helped me to see that those days are over and we can feel a sense of cultural pride in our regalia once again.

During the sewing circle, when the women were gathered inside there was a young man sitting outside the building next to a canvas tent and an open fire. At first I didn't know why he was there, but later I learned that he was designated as the sewing circle's fire keeper. He also was tasked with making sure that the large silver pot of fresh Labrador tea on the table was topped up and warm throughout the day for us. On the last day of the sewing circle, the women invited him into the room and formed a circle around him to surprise him with a gift. Earlier on, one of the women had asked to trace his feet on a piece of brown paper bag, and told him that it was because he looked like he had the same size feet as her son and she needed to use his footprint as a measure. Little did he know that she was tracing his feet to make him his very own pair of moccasins in return for keeping the fire burning for us. When he was presented with the moccasins, he blushed and grinned from ear to ear. The women encouraged him to try them on and they fit him perfectly. Someone started playing jigging music on their phone, and one by one the women started dancing around him. The young fire keeper danced in the middle of the circle until every woman had gotten a turn swinging around arm in arm laughing together. This is what needs to happen in our communities. When Indigenous women come together, we lift everyone up.

Various epidemics wiped out many Indigenous people across the north starting with when the first fur traders arrived in the late sixteenth and seventeenth centuries, bringing with them new sickness. The people were not aware that the blankets they were given by the Hudson Bay Company had poison on them. Paul Mackenzie, the well-known healer from T’èʔehdaà, says that medicine sometimes didn’t work to stave off the disease because the people were blindsided and didn’t have time to prepare. Many of the Dene who died of the flu were buried in very shallow graves and their belongings were burned by the few who managed to survive. Methods like burning spruce tips inside their homes were practiced to kill the disease, says Paul.

Elder Fibbie Tatti, a Sahtúot’įnę Dene woman and former languages commissioner of the NWT, who worked hard to preserve the Dene language and by doing so was instrumental in producing a Dene Kədə́ curriculum representative of all the Indigenous languages, was born in a tent on the banks of the Sahtúdə́, Bear River, in a place called Bennett Field, between Tulita and Délı̨nę, during a tuberculosis outbreak where many people fell ill. Fibbie was raised by her dad and her grandfather. "I was seven years old when I found out that I had a mother." Fibbie's mother was hospitalized with tuberculosis for such a long time that Fibbie did not know she had a mother at all.

There are many people like Fibbie who lost family members to the widespread epidemics that hit the north by storm. Many children did not survive like my grandmother's twin sister. Then there are those who somehow survived against all odds. I once heard about a young child who was the only survivor of the flu epidemic in his entire village. He was only a toddler but somehow he found his way to another village in his canoe. It was there that he was raised by another family for the remainder of his childhood.

The late J.B. Rabesca, an Elder from Łutsël K'é, told his son Alec Rabesca and others the story about his birth and the early years of his life spent out on the land. Alec shared with me the story of his grandmother's strength and survival while I visited him at his home in Łutsël K'é.

I asked Stephanie Poole, the community guide to take me to his house. She drove me there on the back of her four-wheeler. I felt silly bringing my big black leather briefcase with me and I was entirely aware that I looked like a stuffy government worker, but I had bought the bag before starting law school thinking I would need to look professional in court one day. I winced when Stephanie threw my bag with computer inside into a yellow milk crate that was secured to the front of her four-wheeler with a blue nylon rope. It was so bumpy and muddy on the road to Alec's house that I was glad I was still wearing the rubber boots I borrowed from my cousin to go ice fishing.

When I arrived at Alec's house, I was greeted by a husky puppy sitting outside on the steps. Alec pulled back the curtains and looked out the window at his dog and saw me. Once inside, Alec tells me that he has to watch his dog closely because there are wolves around. Alec used to hunt wolves in the tundra and knows all about wolf behavior. He would sometimes get up to thirty pelts at a time, he says nonchalantly. He was one of the last hunters in the area to trap wolves and sell the fur as a form of livelihood.

Alec was born in a small cabin outside of Łutsël K'é. When he was young, Alec's dad used to bring him out on his trapline. Alec learned all there is to know about being out on the land from his father, but he doesn't go out much anymore because he had an operation on his heart, has cataracts in both his eyes, and had his leg amputated below the knee because of diabetes—but he has a cabin a bit out of town that he still goes to when he can. Alec points to a photo of his father in his living room. In the photo his father is wearing a blue sweater and matching blue hat. He is smiling and his hands are open to the sky. He is playing a Dene game called hand games.

Alec tells me that most of his father's immediate family died during the flu epidemic. Alec's grandmother Marie Rabesca was pregnant with his dad when her family died. Both her parents and her six siblings did not make it. Only she and her unborn son survived. Alec believes that she didn't get sick because she was pregnant which gave her a stronger immunity to the disease.

With no other choice but to keep moving forward, Marie began her long journey on foot and walked out of the community pregnant and alone with two dogs at her side. J.B. was born in a small teepee that his mother constructed all on her own, possibly while in active labor.

She gave birth alone in the bush.

Marie and her son J.B. were alone for many years. Most of the time, J.B. sat in a sleigh that his mother had made with the materials around her. When he was old enough, Marie would take one dog with her and follow the shoreline looking for caribou. She would leave the other dog with J.B. and he would keep the fire going. "They had no gun. She killed caribou with a spear," Alec says. If she was successful in getting a caribou, she would make dry meat for her son, herself and the dogs. She also made a caribou-hide parka and pants for her and J.B. so they would be warm.

It wasn't until Alec's father was old enough to remember, about five or six, that they finally saw other people. They met up and joined with another Dene family. Before that, they traveled hundreds of kilometers crossing terrain that most men couldn't fathom surviving. "In the winter in the barren lands everything is white," says Alec. Alec and most other Dene hunters can look at the moon or the sun and tell where they are by the way it is positioned in the sky. The Dene, when traveling across the north, would often look to the stars to map their way home if they were lost, and most skilled hunters still know how to use the stars to navigate instead of relying on GPS or a map.

Elder Jonas Noel, Chief Fred's traveling companion, said his father taught him all that he knew about the sky. He told him that there is a star that only appears every fifty years. Jonas has seen it only once in his lifetime, when he was out on the land. The star Jonas's father was referring to is Sirius, which indeed only appears every fifty years. It goes to show just how closely the Dene were mapping the stars and passing that knowledge down through the centuries. I hope that Jonas gets to see that star again in his lifetime.

There are certain rocks on top of the hills that serve as landmarks. If there are rocks piled up and there is one big flat rock on top, that is a marker pointing in the direction of where to go. These markers are what Alec's grandmother would follow. Alec doesn't use GPS or a paper map, and nor did his grandmother.

Alec tells me that his grandmother lived until she was over a hundred years old. She was a modest, humble woman who lived in a small teepee even in her Elderly years. She didn't want to live in a house in the summer. She preferred to sit outside, where she would have a small fire going.

It is knowledge, like Marie's, of how to survive on the land that saved many communities from being completely eradicated in the epidemics of the past. And now, after Covid, we are facing the largest pandemic yet—global warming. All that we will have at our disposal is our ability to adapt and survive like Marie and J.B.

"When you look at the moon we call her Grandmother, when we look at the sun we call him Grandfather so when you are looking at Grandmother, she is the one that takes care of the water. When the moon is full the tides come in and out, they rise high and they go low, and when that happens there is a tension in all her daughters." This tension, says Gilbert Cazon, husband of Mary Jane Cazon, is the start of a woman's monthly cycle. "Before daughters give birth they give water out of respect. That's the connection that women have to Mother Earth."

Just like our mother is our original home, so too is Mother Earth. The womb is a dam. Released, it brings life force into the world.

My mother had three miscarriages before me, and did not have any more children after she had me. My sister was customarily adopted. How she was able to carry me to full term is a mystery. Some say we choose our mothers. That each one of us has chosen to live on this planet at this exact time in history. As mothers it is our job to teach our children how to live without us. It is the most fulfilling yet heartbreaking role, because we are raising our children to be independent enough to leave us. Have you ever stopped and asked yourself why we are given gifts on our birthday? Instead of receiving gifts we should be giving our parents gifts on the day we were born.

I did not have a natural birth when it came time for my children to come into the world. They were born through cesarean sections, and I feel that the experience of natural birth was taken away from me by the Western healthcare system. The Dene, had celebrations around pregnancy and delivery and still do. We did not need doctors to help women give birth. Now, when a woman has a child, that child is immediately given a number and becomes a statistic. But the Dene had a different way of looking at the birth of a new life outside of having to enter our children's names and information into a system, it was once commonplace to have special naming ceremonies for our children.

Mary Jane Cazon knows how important it is to share the intimate knowledge that was passed down to her to keep our cultural practices alive. Mary Jane is a Dene Zhatıé language specialist who has been working at the school in Łíídlı̨ı̨ Kų́ę́ for the past twenty years. Just like Elder Celine, Mary Jane's father told her that the world was changing and advised her that she needed to be fully educated because one day she would no longer be able to live on the land. Once Mary Jane moved into the community of Łíídlı̨ı̨ Kų́ę́, she was the oldest child in kindergarten. She picked up English as a second language very quickly.

Together, Mary Jane and Gilbert have four children and eight grandchildren. Her mother is Dora Nayally and her father is Boniface Nayally—Náı̨lı̨ meaning "waterfall." Her family is from the Fish Lake and Black Water area in the interior of the Dehcho Region. Her family's camp was located in Black Water. It was there that she was raised on the land until the age of twelve. Mary Jane has a message for young women on the importance of following certain rituals when a young woman gets her first menstrual cycle. She uses her own experience as a teaching tool.

When she first got her period, she had to stay indoors and keep her back against a wall. Her mother told her father that she had started her cycle and right away he went out and collected spruce bough for her. From the entrance of their house through the hallway leading into her room, her parents placed a trail of spruce bough on the floor. It was there in her room that Mary Jane stayed for the duration of her cycle on a soft bed of spruce bough. Her mother told the Elderly women in the community that she had started her first moon, and the women came to see her and shared with her many lessons on how to take care of herself, her home, and how to become a mother to take care of her own children one day. "Each one of them gave me a piece of knowledge on how to become a woman," she recalls.

Then the oldest woman from the community, Elder Liza Yendo, brought Mary Jane material to sew. She already had been given a needle, a pair of scissors, an awl, some beads and uppers from her mother, but she confesses she was still not very good at sewing. Liza asked Mary Jane to sew a pair of moccasins before her next cycle. When Liza left, Mary Jane sewed the moccasins together quickly because she wanted to play outside. As she sewed, her mother made sure to check her work, and found a few knots and loose threads hanging. "I had to undo it seven times and then finally on the eighth time I really took my time" says Mary Jane, and that's when her mother said, "Okay you can go play outside now.'"

Before Mary Jane's next cycle, old lady Liza came back and inspected the moccasins very closely. "She was... checking the beads... making sure there wasn't... any thread hanging... making sure that I did everything well." After she inspected Mary Jane's moccasins, Liza said to her that because she hadn't completed the last few pieces, one day she would have a hill that she would have to climb over, and she would have to learn to live again. "She was telling me about my lifeline right there through my sewing."

"She was telling me about my lifeline right there through my sewing."

And so it was that Mary Jane came to a hill later on in her life that was very difficult for her to climb. She lost one of her children.

The body is a vessel. We only have one body during our lifespan on Mother Earth and we must take care of it by eating healthily, getting lots of rest and enough exercise We are made up of mostly water and so is the earth. Our bodies are a mirrored reflection of Mother Earth. Water is our lifeblood. The rivers and creeks are our veins. Our bones are the rocks. Our skin is the soil. The trees are our lungs. In order to live healthy, long lives, we need Mother Earth to be healthy too. We are healthy when the land and the water are healthy. According to the *Book of Dene,* when Mother Earth was made by the heavens the ground was covered with the hide of a large moose that was soft to the touch. It's no wonder then that our clothes were made out of hide and fur was a second skin. The sewing skills that my grandma and many other Dene people hold sacred, and have honed over time, are also an important part of taking care of the afterbirth of a child.

Mary Jane says that when a child is born, the placenta is supposed to be cared for in a certain way. Her mother taught her how to make a special bag for it. "It has to be a white cloth or canvas, or if you have caribou hide it's even better." Mary Jane's mother told her never to use a regular needle to sew the placenta bag together, but to use an awl instead. The holes must be an inch apart, and the hide must have two layers, and "once you're done you put four young spruce boughs at the bottom to represent the four directions, and then you put a layer of moss." Once that is done the placenta is then to be covered with moss and four more spruce boughs, and then a long piece of caribou sinew is used to tie the awl and sew it through the holes in the hide to keep it loose. It is very important that you don't use a needle, Mary Jane stresses. "Once you put the placenta out on the land, the grandmothers and grandfathers will direct your child." The spirit of the child will go out from the placenta through the loose loopholes that were made—"that's why you don't tighten the bag."

When the bag has been prepared, you must go out and find a grandfather tree—often it is an old poplar tree. It is even better if you can find a grandmother and grandfather tree together, says Mary Jane. Then you place the placenta bag in between the branches of the tree. Once that is done, an offering is made, usually of tobacco, and you talk to the tree and ask it to help raise your child so that the child will be guided by the spirit of their ancestors.

Mary Jane hopes that the young mothers of today can take care of the placentas after childbirth. "When you have your baby in the hospital you should make sure to ask the doctor for the placenta, otherwise it will be thrown out into the garbage and it's not a good thing," she says. "Children need that connection." The connection to the land and their ancestors.

That connection she is referring to is the connection with nature. The placenta is a very sacred part of the birthing process and should be given to Mother Earth. If the placenta is not cared for, it could be seen as dishonoring the vital role it plays in the beginning of one's journey in this life.

A placenta is referred to by the medical industry as a "biochemical waste product." Many women tend not to think about the afterbirth when giving birth inside of a hospital because the doctors and nurses take care of it, but placentas are often stored as a biohazard material before being thrown away. If we would not throw away our scraps and threads without first honoring them, then doctors and nurses should not be allowed to discard the most sacred part of a mother and her bond she has with her infant through pregnancy, the womb, the flesh of our flesh that carries life.

If you put the umbilical cord on the land the same way the placenta is put on the land, then Mother Earth will sing for the child and help them with what they need, says Mary Jane. "You do not tell the child what you want them to be, you just let the child be. Mother Earth will determine what they should do." When you do things in this way, with intention, that connection will always be there between the child, the parents and Mother Earth, and "the child will never be lost." If you place the umbilical cord in the child's home territory, they will have an invisible unbreakable cord connecting them to their birthplace so that they will always have a sense of connection and belonging.

As a child grows up, they slowly begin to relinquish their childhood, says Mary Jane. When they lose their first set of baby teeth the parents are supposed to place the teeth in between the bark of a tree. "Then you dance around... and you ask the weasel to come and help replace the teeth of the child." I wish I had known this when my children were small instead of having to pretend there was such thing as a tooth fairy and placing emphasis yet again on money by putting coins under their pillow following societal norms that are really not that normal at all when you think about it says Mary Jane.

Every day was something new and exciting when Mary Jane was a child. When she was five years old, she recalls going on an excursion for the day with her whole family including her extended family—aunts, uncles, grandma and grandfather. That day they packed up dry meat, animal fat, dry fish and pemmican. Pemmican is a soft dried meat that is pounded into a floury texture, and is often mixed with berries for the Elderly to eat if they don't have any teeth. Her family also brought a pot filled with different plants that they put into a packsack. "I remember we went for such a long walk that day," says Mary Jane. They finally stopped when they came upon a pond where Mary Jane's grandmother put a caribou blanket down on the ground by the shoreline. Mary Jane sat down on the blanket and that is when she was told that she was about to learn a lesson. More specifically, she was about to learn a lesson about how to raise her children.

There was a beaver house nearby, Mary Jane remembers. "We sat there all day waiting for the beavers to come, which was a lesson in patience all on its own. After the mother came out, three little young ones came out." Mary Jane was told to keep a close eye on them. As Mary Jane watched, the mother beaver stood up, making sure there was no danger around before going about her day. The beaver kits followed as she made her way toward a poplar tree. The furry family gathered around it and gnawed on it. Then all of a sudden, a whisky jack started circling the beavers and squawking. "Look across the lake," said one of Mary Jane's family members in a very quiet voice. There, in the trees, was a wolf slinking out of the forest on the hunt. At the same time Mary Jane saw the wolf so did the beaver mother and she dove into the pond and slapped her tail several times. Two of the three kits in her litter followed her into the pond. Then the mother came back up to the surface and saw that the other little one was still gnawing on the bark. She slapped her tail really hard, and the little beaver "rolled down the hill and fell into the water. It just went *plunk*." What Mary Jane saw unfold that day was a lesson she needed to understand. Her mother told her that human beings are like beavers. They take care of their homes and their young ones. They raise their young ones well and they are fierce protectors.

Stacey Sundberg's ehtsı̨ showed her where a small tree is now growing beside the sacred tree that fell shortly after her son was born, which Stacey believes is the mark of a new generation. Stacey has been told by her Elders that our children are coming back reincarnated "to teach us about love again." She says, "We are losing sight of love," and we need children to fill our hearts with love and bring joy. A mother's job is to nurture that joy.

Mothers are the conduit of this reconnection to love, which is why life-givers should be treated with the highest honor, but the long-standing Indian Act—a harmful colonial government tool enacted in 1876 with the goal of assimilation—dispossessed Indigenous women, and caused a patriarchal society in Indigenous communities across Canada by dismantling our social structures in order to render our matriarchs powerless. "We used to have clan mothers that our men Chiefs used to go to before they made any decisions. They were not to ever make a decision before coming to the clan mothers," Stacey explains. The men back then would take the women's advice. The Indian Act tried to abolish women's rights to break us apart, because the women were the ones that held the communities together she adds. "They tried to get rid of us in any which way they can." Now we are left with a misogynistic approach to leadership copied from the kwet'ı̨̀ı̨, white man's way of doing things, where Indigenous women leaders are few and far between, Stacey tells me. We need more women in leadership to make the important decisions, because we think with our hearts.

Stacey is right, Indigenous women were the matriarchs of their community prior to colonization and there needs to be a resurgence—an Indigenous sisterhood, if you will.

Dr. Nicole Redvers, a Dene woman who was born and raised in Deninu Kų́ę́ First Nation, as well in Hay River, is the author of *The Science of the Sacred*, a book about Indigenous medicine, planetary health and earth justice. "The men need to step aside and let the women get to business," she tells me, and even though she is lightheartedly laughing, there is an urgency in her voice. "The feminine aspect to Mother Earth is so strong that we really need women's voices in leadership to come into the circle, and we need men to stand up and step back and honor their women as part of this journey."

Dr. Redvers strongly believes that what's needed for climate solutions to be enacted is enabling more women in leadership positions, because we once lived in a matriarchal system that worked well before colonial disruption.

Melaw Nakehk'o Antoine says that women and men's roles can be a very sensitive topic, depending on how people are raised and how religion has impacted them. Melaw is a northern Dene multidisciplinary artist and hide tanner with lineage in the Dehcho. A descendant of strong leaders in her community, Melaw knows that women once held important leadership roles. They were involved in all decision-making processes, and made a lot of contributions through their advice and through different ways of communicating. "There are stories where people would gather and talk all night long and come to an agreement." There would be a spokesperson, headman—or the Chief, as they're now called—who would relay the decision to the community, and women were a part of that. "Women were always a part of the conversation," she adds.

Melaw was raised to respect protocols within families, and she acknowledges that everyone's family is different. There were no strict gender roles in Melaw's family. When she first started tanning hides, she was told that her grandfather used to help her grandmother tan hides. "My grandfather used to sew his own moccasins. All of the men did. All of the men knew how to tan hides and all of the men knew how to sew, because they would be out on the land for weeks and months at a time, and if something happened they would need to repair their own clothes and their own shoes, or if they needed another hide they needed to know how to prepare that—so everybody knew how to do everything."

On the flip side, Melaw's grandmother was an excellent hunter. "There were so many stories about what a great hunter my grandmother was." Melaw also has an aunt who was a great hunter, who conducted hide-tanning camps in Russia. Melaw has always looked up to her. "She used to be such a badass."

I asked my friend Jennifer Duncan to meet me in the city of Vancouver while I was there for work. After dinner, we made our way down the busy boardwalk until we found a quiet bench to sit on and talk. Jennifer is a Dene lawyer from K'áhbamítúé, Colville Lake, who lives away from home, but still works in her community. She draws on her own experiences as a hunter and fisher, dispelling some of the tropes and stereotypes around the notion that Dene women don't know how to or should not hunt or fish. "We've always been taught to be very independent. To be able to get our own food, haul our own water, get our own firewood. To be able to survive." Even though she is far from home and no longer out on the land as often as she would like, Jennifer would still be able to call on those hunting and fishing skills as needed because it's like riding a bike, you never forget how.

When it comes to women hunting, some of the same questions about gender roles exist around woman drumming. My mother's friend Tanya Lantz has a strong desire to learn how to drum. Tanya is a Dëne Sųłıné woman from Łutsël K'é. She is a mother, daughter, auntie, cousin and friend. Tanya works for the Mackenzie Valley Land and Water Board (MVLWB) and is responsible for engagement, consultation and cultural awareness. As a single parent, Tanya has two roles in her family—the mother and father role—and I can relate to what that feels like, having been a single parent myself.

Forty years ago, her grandmother might have said it was not a good idea for a woman to drum, but today Tanya thinks her grandmother would encourage her to sing drum songs because gender roles have changed dramatically. "I actually bought a drum last week for the first time. I brought it home in a bag." But because the climate is so dry in the north, when she got home and took it out of the bag she noticed a tear. The next day the tear was bigger. That's when Tanya realized the drum wasn't meant for her, so she put it on the shelf. She says she will get a new one when the time is right. Tanya is not letting the belief that Dene women can't drum stop her. Dene women drumming is frowned upon by many people in our northern communities. I was told by several people that the reason women are not supposed to drum is because we are too powerful. I was also told that women are to never step over a man's hunting gun because it would cause him to be unsuccessful when out hunting.

It is in knowing that women are still considered powerful that I find solace and for that reason our power should be celebrated instead of hidden. But the belief by many is that the beating of the drum is in keeping with the rhythm of a heartbeat and because women are life-givers with connection to the spirit world the men are supposed to drum for us as a form of honor; however, I personally feel that women should be able to drum alongside men if they so choose especially in this time of great uncertainty in the world.

"I hope personally one day I get to learn those traditional drum songs and sing them proudly," says Tanya. I hope that for her too and am cheering her on.

Indigenous women have stood outside in the cold on the front lines banging their drums as they are arrested for blocking roads in an attempt to stop industry from trespassing on their lands. The Wet'suwet'en hereditary leaders, our Dene cousins in the south, are known as land defenders for trying to stop a pipeline from running through their territory. They are also considered criminals and under surveillance by the government because of their protests.

We are living in an age where we are in dire need of matriarchal land defenders who are not afraid to stand up for what's right. The words of Anishinaabe environmental activist Winona LaDuke, speaking out about the fight for clean water, sadly still rings true nearly fifty years later: "Someone needs to explain to me why wanting clean drinking water makes you an activist, and why proposing to destroy water with chemical warfare doesn't make a corporation a terrorist." Indigenous youth activists like Autumn Peltier are speaking up for the cause, and giving powerful warnings of what the future might look like if we keep going down the road of prioritizing profit. "We can't eat money, or drink oil."

Man camps pop up in northern Indigenous communities across Canada all the time when male dominated extractive industries move in. With the influx of these man camps, transient industrial workers come in troves and increase violence toward Indigenous women and girls, who in their eyes are seen as worthless. Author Helen Knott, who is part Dane zaa, meaning "those who live among the beaver," as part of the broader Dene group outside of the NWT, has written about the sexual violence occurring at alarming rates in man camps, and how this violence is eerily similar to the raping and pillaging of the land at the hands of the men in man camps.

"With the exploitation of Mother Earth, the majority of that exploitation is done outside of our traditional territory and in some far-off boardroom run by non-Indigenous people—most likely men," says my cousin Cindy Allen, as we sit in my living room visiting over medicinal tea harvested from our home territory. It's patriarchy suppressing matriarchy. Men suppressing women. Men trying to control Mother Earth and her resources. "Mother Earth, though, is not going to put up with it," Cindy exclaims.

Resource extraction is an annoyance to Mother Earth, and global warming is a warning that unless humanity gets our act together and finds a way to live in harmony with her, she can and will wipe us out. She could keep existing, while we might go through another pivotal turning point in history such as the ice age except this time it will be by fire and drought that would make it unbearable for us. "She has it within her power to right things for herself, but that might not include us human beings in the future." Cindy takes time to choose her words wisely. "That would be her justice. She doesn't have to take the harm against her quietly, and I don't think she is right now."

"She doesn't have to take the harm against her quietly."

Cindy is the granddaughter of Gabriel Doctor and Mary Adele Doctor. Her late mother is Christine Doctor Allen. Her father, Richard Allen, worked for the Hudson's Bay Company. Cindy is the great-great-granddaughter of Chief Monfwi, the Tłı̨chǫ Chief signatory to Treaty 11. I love Cindy's style. Every time I see her, she looks like a piece of art. She has on an embroidered Dene flower vest, and running shoes with red laces and a wolf design by a Kwakwaka'wakw artist. I compliment her outfit as she sits across from me. Cindy and I are among only a handful of Yellowknives Dene members who have graduated with a law degree.

"It's very disturbing for me to think about fracking because what that is, is you're blowing up Mother Earth from the inside," Cindy remarks. It is so harmful—these "explosions inside Mother Earth." Extracting resources only to leave a toxic legacy. She gives a big sigh. "It's not a good feeling and not a good future to look forward to."

Cindy hopes that our Dene stories, teachings and truths from Indigenous voices and knowledge carriers will be heeded so that everyone can one day walk a path of harmony that upholds Indigenous communities and Mother Earth at the same time.

Cindy tells me she would like to one day earn her Doctor name literally. "I want to get a PhD on Dene Law and stories and work with women." Dene stories contain our laws and teach us how to live. "If you hold those stories true, close to your heart and keep them in your mind, they'll help guide you into the future."

A few years ago, Cindy was visiting the North Peace Cultural Centre in Fort St. John, BC, and she saw a large painting of what she believes to be Yamǫǫ̀zha's beaver wife on the ceiling.

Cindy has done her research, and has her own interpretation of the legend of Yamǫǫ̀zha and his beaver wife. In the version Fred Sangris tells, the wife is still alive until she is solidified in the stars next to Yamǫǫ̀zha with her shawl. Cindy's interpretation has been published in *Reclaiming Power and Place: Final Report of the National Inquiry into Missing and Murdered Indigenous Women and Girls,* a scathing indictment of the government and RCMP's failure to do enough to protect Indigenous women and girls.

The MMIWG report came about after there was cause for concern about the ongoing extreme rates of violence toward Indigenous women in Canada due to their vulnerability as a demographic—beginning with the Indian Act—making them highly susceptible to predatory behavior by men, primarily white men, who believe Indigenous women and girls are disposable. These men have been able to get away with the abuse of Indigenous women and girls because of the lax justice system, which historically has not brought down fair disciplinary actions.

Cindy has in her hand a children's book and is turning the pages. The book is called *Yamozha and His Beaver Wife.* She flips through the pages which contain dramatic illustrations by Archie Beaulieu, a renowned Tłı̨chǫ artist whose work is also showcased on the cover of this book. The book was published with the support of the Yellowknife Catholic School District. It is just one of many versions of the same story, which speaks to the marriage agreement between Yamǫǫ̀zha and his beaver wife. The story has been lost in translation. The colorful artwork is distracting, making the subject matter seem light even though the story itself is foreboding.

"He came across her out on the land and she agreed to marry him so long as she doesn't get her feet wet in grassy water or when she's going over little creeks," says Cindy. When he meets his soon-to-be-wife, she is human and can't get her feet wet or she will turn back into a beaver, but he does not know this. "For her to be human she had to keep dry, just like a mermaid." To me it almost sounds like a play on the old adage "cold feet" before a wedding.

Yamǫǫ̀zha laughs when his beaver wife tells him she cannot risk getting her feet wet, and tells her she doesn't need to worry. In the book he says, "I will take very good care of you." In his mind, it is an easy task to do as she has asked. They live for a long time traveling the land until things take a turn for the worse.

At one point the couple come to a small creek and Yamǫǫ̀zha decides not to put willows or branches down for his wife to cross the water, because he thinks it is easy enough for her to cross. Yamǫǫ̀zha can "walk hundreds of feet in a single step," so it is very easy for him to cross but he forgets that it is not easy for his wife. When he doesn't put down the willow for her, she changes back into a beaver and leaves him. It is supposed to be a love story but it doesn't sound like one, says Cindy. "It's very violent and has a tragic, sad violent ending." There is no happy-ever-after.

When she does not follow him, he becomes angry. But the man is supposed to chart a path forward that is safe, says Cindy. "It's the responsibility of the man to break trail." Yamǫǫ̀zha's responsibility was to lay down the branches so that his wife, who is carrying their camp on her back, has a safe path to walk upon. Dene people lived a long time in a harsh environment in the north, and it took the man and the woman working together to survive she cries.

Cindy is clearly upset as she summarizes the story. Yamǫǫ̀zha is so angry that he chases his wife all over the land.

He finally catches up to her where she has built a beaver lodge, and he digs into it. He takes the baby beaver from inside and he kills it, then cooks it and eats it. "Tragedy for the beaver wife. Whose baby is that? He killed his own baby and ate it. Cannibalizing his own kin." The thought of this travesty taking place deeply troubles Cindy and she has to pause for a moment. She takes a deep breath. "There's other ways it could have been handled," she says.

Yamǫ̀ǫ̀zha's wife almost escapes him; she goes up the Dehcho and is just about to reach the Arctic Ocean, but he catches up to her and transforms her into a rock. "He could not stand the fact that she had enough power to leave him and that she could escape." He doesn't want to let her leave, but the marriage promise has been broken and she has the right to go.

As I listen again to the story and apply all of what I've been researching, I veer away from the domestic aspects of the story for a moment and wonder if Yamǫ̀ǫ̀zha going to the end of the land toward the Arctic Ocean is a sign of things to come. Considering the present-day race to the Arctic Ocean for the last of the world's oil, industry will need access to transport it, and it just so happens that a road from Iñuuvik to Tuktuuyaqtuuq has been built with speculation that it is for that very purpose. Once again pillaging and raping mother earth to extract and exploit.

"It's a hard story to hear," says Cindy, pointing out the fact that the male perspective is there but the woman's voice is silenced. "When I think about the story in relation to what the beaver wife had to endure, women today are enduring similar things—including Mother Earth, she is also enduring a lot of harm. Silence of the violence means you condone it. Things will not get better for our communities and our women and girls if we keep silent and don't tell the truth, even though it might be hard and challenging to speak that truth. We have to honor our moms, we have to honor our grandmothers, we have to honor our girls."

There was a marriage agreement, says Cindy, and there are obligations on the woman and the man in any marriage agreement. If Yamǫǫ̀zha had fulfilled his promise to his wife, there would have been no conflict and they would have been able to live in harmony. Yamǫǫ̀zha is the Dene lawmaker, Cindy explains, he was a cultural hero who had a lot of medicine power. He was a transformer. There are stories of him transforming into other animals including a beaver, and living with a family of beavers happily for a long time so he could understand the way that they live. "He could have transformed himself into a beaver and regained her trust so that she felt that she could be human again with him in safety and live a good life together. He could have done that. He chose not to." Instead, he let his emotions take over. So while he's our Dene lawmaker, he also has faults and needs guidance and help—because there are other choices he could have made instead of acting out of violence, rage and retribution. He could have taken the peacemaker route, and reconciled with his wife. Today, the workers in the man camps can turn away and make peace with Mother Earth.

The most widely known interpretation of the story of Yamǫǫ̀zha and his beaver wife is that Yamǫǫ̀zha made the land safe by chasing giant beavers—and left his mark on the landscapes across the north, from Wıìlıìdeh to the Sahtúdə́ rapids, the ramparts on the Dehcho, and the rock near the Arctic Ocean. There's also a place in Tulita known as the "ever burning fire," where the grease of the beavers he killed dripped as he was cooking. Smoke can always be seen there; it too is a sacred site.

George Blondin's book *Yamoria the Lawmaker* includes the Sahtú version of the story of the beaver wife, though it centers Yamǫǫ̀zha's brother Yamǫ́rıa. In it, Yamǫ́rıa is tricked by a woman who is a giant beaver. They live together for a long time, and she tells him to always put a willow down to cross a creek and emphasizes the importance of doing so. Yamǫ́rıa doesn't understand why it is so important but he agrees. One time he forgets to put the willow down and crosses the creek, and when he returns to their home she is gone. That's when he finds out she is a beaver and he is angry that he has been tricked. He chases the family of Tsàcho and kills them, placing their hides on Bear Rock. Could it be that the beaver wife was actually not a good character in the story after all? There are always two sides to every story and we don't have her side because she has not been given the opportunity to tell her side of the story.

Cindy and I went for a walk after the traumatic reading of the story of Yamǫǫ̀zha and his beaver wife. As we were walking, we came across a dead black squirrel on the sidewalk in front of us. At first, I didn't notice what it was, and I walked over it. But Cindy stopped in her tracks. "It's a squirrel," she exclaimed. She didn't want to pass it by without giving it a proper burial, so she picked it up and put it under a nearby tree. She always carries with her a pouch of tobacco, and she took a pinch of it and put it down beside the squirrel and said a prayer.

Squirrels die all the time, deer and racoons get hit by cars in the south, and I have to admit I have become somewhat desensitized to seeing roadkill. But Cindy tells me that because it was in our path it could be an omen. How would one of us like it if we had our insides ripped out, only to be discarded and left to dry in the sun? Cindy is the type of person who saves mice when her cat comes home with them. She is as tiny as a mouse herself.

Imagine if every single one of us took the time to take care of the spirit of the animals like Cindy. What if we all recognized that no matter how big or small, dead or alive, animals should be treated with dignity.

— 4 —

Spirit of the Animals

In many Dene stories it is animals who are saving the day, not humans. Limited scientific beliefs such as the theory of evolution and Darwin's natural selection, which claim that humans are above all other life on earth, place us in competition with the ecosystem rather than in connection with it. Darwin has given man the excuse to dominate and destroy nature for personal gain, but his theories are far from Dene Law, where not only is caring for Mother Earth's creatures at the center of everything we do, we know we also cannot live without them.

Many Indigenous nations across Canada have clan systems. Some families have the spirit of the eagle, some have wolf, wolverine, bear, fox, raven and so on. It is said that those who have the spirit of a certain animal should not harvest that animal. John B. Zoe is a well sought-after Tłı̨chǫ knowledge carrier who lives in Behchokǫ̀. I first came to know John B. when I reached out to him to ask if he would provide feedback on my first novel which was inspired by our Dene stories. In talking about the clan system he says the term "clan" is an English word, and that when we refer to a "clan," what we are really saying is that we are related to a certain animal. Some Dene people will say they don't eat beaver, which is an indication that they are from the beaver clan. Others cannot be in the presence of a wolf that's been harvested. I recall my friend Steven Nitah from Łutsël K'é saying that his grandmother told him he belongs to the wolf clan. "She would get really, really bad headaches if one of her sons or grandsons killed a wolf." Others can't eat bear meat or do anything to harm a bear because they are related, says John B.

This is something that each family quantifies for themselves, he says. Prior to religion coming into our communities, many Dene people were named after the traits of animals, but when the church started writing down names that all changed, says John B. For instance, someone might have had the name "Martin Face," and it would have been translated to something biblical or just shortened to Martin. Even though it may no longer be as obvious, the clan systems are still there. My best friend Jennie McPherson is Sahtúot'įnę, and she tells me that her family has the characteristics of the Ts'ágot'įnę. Breaking down each part of this word you can see the word *ts'a,* which means "beaver," and *got'įnę,* which means "people." Like the beaver, her family is "smart and hardworking," she says proudly. Jennie is certainly no stranger to hard work. I saw this when I stayed with her and her husband James for a few days at their cabin on Great Bear Lake. She put me to work. When we were packing up to leave she made sure that the floors and counters were spotless, so that no mice or bears would get into the cabin while they were away for the winter.

Then there is Mary Jane Cazon, who is very knowledgeable about the clan her ancestors identified with prior to colonial interruption. She is a descendant of the fish clan people on her mother's side.

Elder Lawrence Casaway, who is originally from a place called Rocher River, now lives along the T'èʔehdaà highway. Like Steven Nitah, he believes he has the spirit of the wolf because his grandfather told him not to ever shoot wolf. When I drive up to his house to visit, there are two large dogs guarding the house in the gated dog lot littered with empty doghouses. Lawrence used to have eighty sled dogs in his front yard, and not once did any wolves come around and bother them, but his next-door neighbor had such a terrible wolf problem that his dogs were killed by wolves.

Surrounded by homemade traps leaning against the side of his house, Lawrence knows better than to overharvest. Beavers hold back water for hunters, he says. If it weren't for the beavers holding back water with their dams, the ice would flood and wipe out other animals' homes—like the neighboring muskrat living in push-ups on the surface of the ice. Every animal has a purpose, but the beaver is the hardest worker in the animal kingdom. They are very clean as well. They have two runs, says Lawrence—two routes that they use to enter and exit their house. A front door and a back door. The front door is used for coming and going, and the back door is used as a garbage chute where they dispose of their waste.

Lawrence is a seasoned trapper and because of this he knows which door is which when he is lowering his traps into the ice in front of the dams. Lawrence was also a professional dog sledder in his younger days. He has many awards and medals from mushing dog teams. Although he has won first place many times, he is still modest and doesn't consider himself an expert in either—but I do.

Lawrence has trapped many different animals in his lifetime. Lynx, fox, rabbit, wolverine. He stops at wolverine and tells me that the wolverine's wife is the tamarack tree. His grandfather told him that a wolverine will destroy any other tree around if it is trapped, but it will not even so much as leave a scratch on the tamarack tree because that is the wolverine's wife. This is why Lawrence uses the tamarack tree as a toggle trap for the wolverine and is successful in trapping the wild species every time.

Lawrence does not eat wolverine because it tastes too strong, he says. He only traps wolverine to sell the pelts. He usually gets about five or six wolverines per year but this past winter he only got squirrels because the animals have been scarce. He also hasn't trapped as much because he is getting older.

Lawrence has seen two white muskrats in his lifetime. One was swimming downstream at Wıìlıìdeh. And the other was when he was moose hunting on the Taltson River—in 1974, to be precise. He was sitting on the banks of the river when he saw a trail of bubbles in the water going upstream. He followed the bubbles until he saw something white pop up above the surface. He shot it with his .22 rifle. He shows me the white muskrat fur that is framed and mounted on his wall. It's as white as his mustache and white hair under his baseball cap that reads "Bathurst Caribou." He is sitting next to a small white Christmas tree on his porch as we visit. There is a dream catcher behind him on the side of the house next to an old wreath, and for a moment he looks like Santa Claus on vacation as the hummingbird feeders above my head sway in the soft summer wind.

Out of all the animals he traps, his favorite is the beaver. Lawrence goes out and checks his traps every other day when he has them set. He explains to me in detail how he traps them, and I frantically scribble down his words in my notebook with a cheap pen that's just about to run out of ink. He prefers that I don't record our conversation. Like many Elders it makes him uncomfortable. Even taking notes while an Elder is talking is seen as a bit of an interference, but Lawrence lets me scribble away when I tell him I don't have a very good memory and I don't want to miss anything important, which is basically everything he is saying. There's something magical that happens when nothing is being recorded, and it's the same thing that happens off the record which is why sometimes it's better to capture a moment in real time rather than hitting record.

Lawrence spent many years teaching kids about life on the land. Once, while teaching a class about trapping, Lawrence says a young person really put him to the test. The student asked how he had lived so long by only eating wild harvested foods, which is a diet of little to no vegetables. "I really had to think about that one." Assuming the child was trying to make a case to his parents for not having to eat his fruits and vegetables, Lawrence answered: "Before you ate that rabbit, it was eating branches." A rabbit is a vegetarian. When we eat wild foods like the rabbit, we are also eating plants and medicines that are good for us. What Lawrence didn't explain to the student was that pollutants and extreme forest fires are starting to impact the animals' ability to harvest berries, plants and medicines. It's something that we know is becoming more and more prevalent but also something that a lot of us live in denial about in the hopes that it might go away on its own. "Without animals we would never survive," Lawrence tells me. The animals are so intelligent that when they are sick they know what plants to eat to make them better.

When teaching students about his trapping methods, Elder Lawrence says that trapping starts at home. I am now a bonified student of Lawrence's as he teaches me the basics of trapping. He first makes his own snare ladder out of sturdy long branches. He nails these together in his work shed where dozens of traps are hanging on the walls next to a few pictures that he's picked up from the dump, among other interesting gadgets that I wouldn't know what to do with. He then chisels a hole in the ice above one of the beaver runs outside the dam, and lowers the ladder into the hole. He knows when he has hit a beaver run when the ice sounds hollow. He pounds the ice with a stick all around the dam until he hears the hollow sound. He says that some people use an ice auger instead of a handheld chisel to get the job done, but he is against it. It's loud and obnoxious and it causes air to come out of the water. Nowadays people also use steel traps, but he doesn't like to use steel either. The metal is too heavy, among other issues. The wooden snare ladder is much lighter he tells me.

The beavers scurry out as soon as they hear any sound and return after a day or so, says Lawrence. By then he has his ladder lowered in the water in the run with stuffed branches on either side so the beavers can't dart around it. The snares are tied to each rung of the ladder, and he packs grass around it to camouflage it. He knows the typical daily routine of the beaver. Like us, they like to sleep in. They don't come out until around just after midday. When he is not trapping beaver, he sometimes shoots them, but "when you shoot a beaver you've got to be quick or they'll sink," he says. Lawrence shows me the hook, called a "bow," that he uses to grab the beaver from afar before it sinks. It looks like a very long and sturdy fish-hook on a straight tree branch. He has it reinforced with rope and ties another stick to it for extra length if the beaver is further away. He uses this hook to grab the beaver from underneath. He emphasizes that you must grab the beaver from underneath not above, or it will sink.

In his younger days Lawrence used to go out in a canvas canoe to go hunting and trapping. Sometimes he would stay out all night and sleep under the stars. He would cook beaver tail, beaver stomach and beaver liver over the fire—"what a meal," he tells me. Personally, I don't enjoy the taste of beaver but he almost has me convinced that it's delicious. It's a lot of work, says Lawrence—you have to skin the beaver and stretch the hide soon after you harvest a beaver, otherwise it will rot. He would often go out alone to check his traps and hunt but sometimes he would go with others. The sun is warm on my back as we chat. I wonder if he has beaver meat in the green fridge at the end of the porch.

"When I take a holiday, my holiday is out in the bush," says Lawrence. He has some red paint splattered on his black and white checkered button-up shirt, and I suspect it might be beaver blood. Lawrence taught culture camps to children and college students for many years. He says there is such thing as an Elder succession plan. Young people need to learn from their Elders so that when they are Elders themselves they can teach the young, and it continues in a cycle just like the cycle of life.

He has a sign in his window that says "Lawrence's classroom" on a piece of wood that one of his students gifted him. Lawrence teaches the importance of respect—respect for each other, the land and animals—and believes it is of the utmost importance. "Young people need respect for one another." When hunters travel together they must respect each other's harvesting areas, he tells me. What he means by this is that if two hunters are on the opposite sides of the same lake, and a beaver appears on one side of the lake, it is for the hunter on that side of the lake to harvest. There can be no race to the animal. One hunter has to stand down. If the hunt is successful, says Lawrence, then "we eat together."

Sharing occurs especially if one hunter is successful and the other isn't. The successful hunter always shares his harvest. The other protocol that Lawrence teaches is that when a lot of people are hunting together, there should be only one communal fire. You can't have one person across the way with their own fire going. It is disrespectful, because the optics are that the hunter with the lone fire doesn't want to share even if that's not what their intention is.

Many non-Indigenous people have a misinformed understanding of what it means to hunt, trap and fish as a form of livelihood. Right now, it is still healthy to live off the land as long as you don't harvest in areas adjacent to human population. The meat that is harvested is healthier than processed foods because the animals aren't exposed to growth hormones and different chemicals. Food processing in a facility where livestock are living in close quarters with no room to move is disrespectful to the animal.

Harvesting local meat is healthier than eating fast food, but unfortunately land animals are now starting to consume and absorb more and more metals and chemicals. Dene academic, Dr. Nicole Redvers, says that when humans eat some parts of certain wild game bioaccumulation can occur, causing a "higher concentration of metals in our system from traditional foods than we have had in the past." If things don't change, this will create challenges for those who live primarily off locally harvested food. The north is waving a "white flag," says Dr. Redvers, and we need to inform others of the changes that are occurring so that there is an awareness of what's coming if we don't do something about it. We need to listen to what the animals are saying and relay to the world the message the animals are telling us. We have to speak for them now. They are crying out.

Randy Baillargeon is one of those hunters who is raising the alarm. I ask Randy about his thoughts on the health of the geese, knowing that he would have some insight into any changes in their diet. Every spring Randy goes geese hunting. "That's my favorite time of the year," he says. Those who know Randy have come to expect to receive a bird from him. The most geese he has ever gotten at one time was when he was with his cousins. Together they shot a hundred. "That was a lot of bird," he says.

Geese are a really unique bird, and the meat has a very rich oily taste that's best cooked over an open fire. But Randy informs me that nowadays when birds come from the south after having eaten a lot of farm food and city food, it's hard to know if they are healthy. When the geese lay their eggs in the barren lands in the north they are ingesting their natural diet, but when they leave for the south and return he's not sure what they are bringing back with them. Randy admits it can be scary sometimes to eat goose.

Most of the time Randy can tell if a bird is sick just by inspecting it. If it has pus or a lump, that's a sure sign that it's a sick bird and he will give it back to the land. "Even though I like hunting them… sometimes I get worried." When giving geese away to community members, he warns people to cook the bird until it is very well done. Randy has yet to see a really sick goose with a stomach full of garbage, but sometimes he accidentally shoots ones that are tagged. The tag is a tracker that shows where the bird has traveled from. One of the ducks he shot down had come all the way from Newfoundland. "It's crazy how far they fly." The tagged birds are usually ones that are sick or have been exposed to oil fields, he explains.

It's safe to say he most likely does not carry the spirit of any winged ones as his spirit animal. Randy remembers when he was young, he used to harvest the seagull eggs on the land around his home, but just as he is cautious about geese he is very wary of eating seagull eggs. Seagulls have a highly acidic digestive system and will regurgitate the plastic garbage materials that they consume but studies have shown that seagulls are flying around with a lot of garbage in their guts.

Dene Author George Blondin writes about the power of seagull medicine in *Yamoria the Lawmaker*, describing how some people with seagull medicine would cut sickness out of people's stomachs, or they would use their own saliva as an ointment or have people drink their saliva to get better. My grandmother always said that seagulls were her favorite. I used to think that maybe it was because seeing them signified that the weather was warming and it was finally summer, and maybe that is true, but I'm also starting to wonder if she knew the power of seagull medicine. Now that pollution has become more prevalent, Randy is right to be cautious about consuming seagull eggs in his diet, but he holds dear the times that his grandpa, Alfred Baillargeon, used to send him out to get fresh eggs for breakfast when he was young. Instead of going to the store or out to a chicken coop on a farm, Randy would look for seagull nests in the wild. In Randy's world there is no fenced-in backyard, just endless miles of wide-open space—both land and water. As a young boy, if he was successful at finding a nest without being attacked, he and his grandpa would sit down to a nice breakfast of fresh seagull eggs. Randy feels sad that he can't make those same memories with his own son. This practice has become a special memory—something he shared with his grandfather. Those breakfasts have become something that he longs for. The only time he collects seagull eggs now, if at all, is at the East Arm of Tinde'e or even further away, because most birds now are city birds and "you just don't know what the birds eat and if that egg is healthy to eat."

Randy reminds me that, whether we know it or not, or accept it, we are already collectively grieving the special shared moments that are a part of the overall cultural lifestyle of our Dene identity, and this amounts to a freedom that we can no longer say we truly have.

Bears are not to be messed with just ask Lee Mandeville, the son of Elder George Mandeville.

When I ask gun-at-the-ready wildlife officer Lee about his wildest stories, he says, hands down, it's dealing with bears. One fall morning, Lee had just finished a course on how to manage wildlife attacks on humans when he got an emergency call. He was just pouring his first coffee of the day when a call came in from the RCMP on the satellite phone, and told him that there was a possible grizzly attack on an American couple in Henry River, close to where Lee's great-great-grandfather Le Camarade du Mandeville had lived. Henry River is an old hunting area for Lee's family inside of the Thelon Wildlife Sanctuary so he knows the area well.

"I was thinking of breakfast on my way to the helicopter," Lee recalls, but that morning he had to skip the most important meal of the day because in his gut Lee knew there was no time to waste. He grabbed an on-duty officer to come with him who he says was inexperienced but capable. The RCMP wanted to accompany Lee in the helicopter, but Lee asked them to stand down because they needed the space for the couple and their gear.

Lee and his small crew flew to Tacheè, where they had to stop to fuel up, and then went on their way again following the coordinates to where the couple were said to have been. "We were coming over a hill, into a river valley, then we could see the rapids, then a camp appearing. We got closer and could see a grizzly bear in attack mode pursuing this couple." It was just like out of a movie, Lee says. "The husband was holding his wife who was buried into his chest and it's the end for them, they've got the rapids behind them, no firearms, no boat, and this bear is pursuing them." The pilot steered the helicopter and flew a few feet over the animal to haze it, scaring it into a large rock crevice. They soon managed to land the helicopter and rescue the couple but Lee's job was still not finished.

Lee got the pilot to take the door off the helicopter and the other armed officer stayed with the couple to keep them safe. Lee didn't want to pursue the animal on foot, so he went back up in the chopper and the pilot hazed it again before Lee took the shot. He didn't miss. The way George, Lee's dad, tells the story makes Lee sound like Rambo.

In the aftermath of the near attack, the couples fold-up canoe was crunched in half. "It ate all their fish. Took their food. It was going back and forth to their camp grabbing some food and then moving away." When all their rations were gone, they were next. The husband had one can of bear spray but he had used it all up and it most likely only served to anger the bear. They had nothing left to defend themselves with, Lee explains. "They were up for twenty-four hours straight with no tent. The tent was ripped up. All they could do was stand there, exhausted and starving."

That same bear had been a problem in the area for three years, says Lee. Prior to the couple's near-death encounter, a group of kids were dropped off in boats at the same location and the bear took over their camp. "All they could do was just sit in the water and watch it in their canoes." Other tourists had been pushed out by the bear, too. It had certainly been marking its territory. There was even an officer who'd tried to relocate it to no avail, and because many people used that area as a camping spot the bear had learned that it was a good place to go to for an easy meal—especially in the fall when it was trying to fatten up for the winter.

That encounter was not Lee's first notable bear story—he has many. Lee was once brought in to deal with a black bear on the all-season Whatì Road. The bear was terrorizing one of the forestry firefighter camps. "They tried everything to deter it. Axes. Chainsaws. Screaming at it. Throwing things at it and it just wouldn't go away. They flew me in there to take care of it." Lee was all alone in the camp while everyone went to go fight fires, and he says it was a bit of an eerie feeling sometimes. He just had to sit there and wait for the animal to come out. "They're so sneaky. They're so quiet." Black bears are notorious for stalking behavior, where they will watch their prey from afar for long periods of time before going in for the kill.

Lee waited three days in the camp, and finally on the fourth day at five o'clock in the morning he could hear one of the crew members banging a pot, which was the signal that the bear was back. Lee was ready for it. He ran out in his underwear with his gun and gave another gun to one of the firefighters, and together they shot the bear. "That bear was pulling on a teenager's foot. He grabbed onto his foot. He's lucky he had his boots on. He was dragging him out of his tent and we managed to shoot it like that." A close call indeed and something out of a scary movie for sure. Bear attacks are more common than one would think in the north. Perhaps it is because there is a bear every hundred square kilometers.

The most terrifying bear encounter Lee ever had was when he was called in by the RCMP to help locate the victim of a grizzly bear attack in Tulita. "That was a game changer for me," he says in a more serious tone. "We flew up in a charter plane with six RCMP officers. It was not a good feeling right from the start." The weather was somber. The clouds were dark. It was windy, and the rain was falling every which way.

When they landed, Lee did a briefing with the officers and they came up with a plan to find the victim of the attack. They began tracking signs of both animal and man in an uphill area. "We were shoulder to shoulder moving through the bush through the thorns and you could see all kinds of evidence, clothing, you could see a struggle all throughout." It was a gory scene, says Lee. He could see where the victim had been buried in the bear's den and then brought out again. Before then he had hoped that the man was alive, he and the rescuers had been calling out his name in vain.

"I could see the trees moving and could hear an animal running through the bush and grunting. A grizzly bear running through the bush at full speed trying to kill you is the scariest feeling. It's do-or-die."

There is a long pause, as Lee holds back tears.

It was the first time he couldn't save someone, he tells me. "It was a very sad ordeal." He has to stop again for a moment, his voice shaking. He coughs to regain his composure, apologizes and continues. "I could see the trees moving and could hear an animal running through the bush and grunting. A grizzly bear running through the bush at full speed trying to kill you is the scariest feeling. It's do-or-die." All the training he had done in his whole life had led up to that moment—and now his life depended on it. The trees were swaying, and Lee could hear the bear charging toward them. One of the officers said it was coming from a different direction, but Lee had the sense that it was on his right, so when it charged at them everyone turned and fired and the bear landed a mere few meters from Lee's feet. "Locking eyes with a grizzly bear just before it's trying to kill you is something else."

One of the officers lost his hearing from the sound of Lee's gun and Lee feels guilty about that but he feels better knowing that the victim's family has closure, and that his body was sent home for a proper burial. "They don't have to live wondering what happened to him." The man was a tourist from another country.

"This is not a job you can just fall into. You have to have lifelong experience." Lee thanks his dad for buying him rounds and rounds of ammunition and letting him do target practice as a kid. A gun is not a weapon for Lee, "it's a tool."

My grandma instilled in me the need to be cautious of bears, and my distant cousin Tommy Lafferty was taught the same.

Tommy was born in Yellowknife but was raised and lives in Behchokǫ̀. He is a Tłı̨chǫ citizen and is still proud of his Métis heritage as well. "We were always taught to respect animals, especially the predators, we always had to be wary of them," he says. If we are ever in a situation where we encounter a large animal when out on the land, "we are supposed to talk to them and tell them we are not there for them and for them to go somewhere else."

So when Tommy encountered a grizzly, that was exactly what he did. He is one of the few people in the world who have lived to tell the tale of surviving a grizzly attack. He has recounted the story many times, but he does not want the details of the incident to be written down on paper—because he promised his good friend, who was also an author, that he could be the one to write it. Sadly his friend got sick and passed away before he could do so, but Tommy has promised to honor his friend by not allowing his story to be published. What he will share is the circumstances around the incident. "I've been to a lot of training courses, bear aware, predator awareness… growing up on the land… we are always around predators so we just have to be mindful of them and take precautions." The type of precautions Tommy is talking about are making sure you have a gun and know how to shoot it, carrying flares, bear bangers or bear spray, "make sure the bear spray is not faulty, and you don't know if it's faulty until you pull that pin."

Tommy was working up at the Colomac Mine Remediation Project at the edge of the treeline and tundra when the bear attack happened that changed his life forever. He had just finished his contract earlier that morning and was supposed to be flying out the next day, but was asked to assist a hydrologist to take water samples around the mine site. "I could have said no and maybe the job would have never got done, maybe he would have went with someone else, maybe something else would have happened. There's a whole lot of maybes."

When I ask him if he feels that he has a spiritual connection to the bear that attacked him he replies in his baritone voice, "For about a year or so before the accident happened, I had dreams of a big bear chasing me. Every time, I dreamed the bear got closer and closer and closer. My father was in my dream." The reoccurring dream, says Tommy, would always start off the same, and near the end his dad would be standing at the truck ready and aiming a gun. It didn't feel like Tommy could ever get closer to his dad as he ran from the bear. He would run in a zigzag so that his dad could get the shot. Every time Tommy veered off from the bear's direct path he'd yell at his dad to shoot, but the shot would never go off—maybe because his father was afraid to miss or accidentally shoot Tommy. The last time Tommy had that dream, it was a few months before the incident, he can't remember exactly when, but the dream was finally complete. "The last thing I remember is that bear finally caught up to me."

It has been twenty years since the attack, but it still feels like yesterday for Tommy. "My healing is laughter." Tommy found the ability to laugh even when in the hospital and he says that is what helped in his recovery. When he woke from surgery he had many specialists around him. "I had all the other doctors, the ortho guy, the neuro guy because of all the damage on my head, and the plastic surgeon coming around."

I saw Tommy in the hospital when he was about to get skin graft surgery. In order for him to get discharged he had to go through a psychiatric assessment. "The psychiatrist said after our meeting... 'Tom, I think that I'm more messed up from your story than you are.'" Tommy laughs, but in all seriousness he says, "Later on, the PTSD bothered me. It was silent. It came out at inopportune times."

Some people would be afraid to go back out on the land after a bear attack, but Tommy had no qualms about getting back to what he loves—which is being outdoors. Seven weeks later, Tommy was doing culvert inspections on the side of the highway. He had been given full clearance to go back to work even though he still had fifty staples in his head and stitches in his face and arm. He'd requested to come back to work because he felt that he needed to work for his own mental health and to provide for his family. "Everyone wanted me to succeed. We all need to succeed. Whether it be being able to make our bed in the morning, to coming home and having a good meal." Tommy says he couldn't just sit around. "My necessity to work and be back on the land was probably the best choice I made to heal. I'm still healing today."

There's not a lot of flashbacks that pop up for Tommy, and he has never let that day overtake his spiritual and mental capabilities. "The anniversary of that day used to be difficult," he explains, but he says it's getting better. The attack happened a week before his middle daughter's birthday. "I was able to make it for my baby's birthday."

As long as I've known Tommy he has always been a good storyteller. When all of us cousins used to visit Nı̨hshìi in the summers he was always telling stories around the fire.

Now he has one more story to add—one of which very few can say they have lived to tell the tale.

Sometimes bears take lives, but on the rare occasion they save them. I believe there was a time that a bear saved my family. My kids and I were driving in the mountains, returning south for my second year of law school after a summer spent back home in Yellowknife. We were going along at a good speed. It was early morning, the fog was just lifting and I could tell it was going to be a nice clear day. I'm always more cautious when driving in the mountains, especially when the weather is not good. The kids were in the back seat on their devices, and I was up front listening to music and drinking my coffee, when I saw a black bear come up over the left side of the ditch. It was clearly limping on its right front paw as it ran across the highway in front of us. When I slammed on the brakes, the kids looked up from their devices. We could see that the bear was not quite an adult bear and not quite a cub, it was somewhere in between. Thankfully no other vehicles were behind us or we would have been rear ended.

The bear crossed the road and ran into the ditch on the other side until it was out of sight in the trees. I kept driving, but this time I didn't go as fast because I was worried about seeing more wildlife. It wasn't long until a large semitruck caught up to our vehicle and was tailing us. Some of the big rigs on the highway can be real bullies. I wanted him to pass me but there was no turning lane on the winding road. I felt the pressure to speed up, but stayed the limit. After about ten minutes of the truck following too closely behind us, I noticed out of the corner of my eye on the left side of the road another black bear—not quite a cub and not quite an adult. The bear crossed in front of us on the highway and was limping on its right paw before disappearing into the ditch on the other side. I looked at the kids to see that they were just as confused as I was. It was identical to the first bear. How could this be possible?

I've often thought about that moment, and the only explanation I can come up with is that the first bear sighting was a warning, to make sure that I slowed down before the semi-truck began following closely behind us, because if I hadn't seen the bear the first time I wouldn't have slowed down and I'd have had to slam on my brakes when seeing the second bear—which would have caused the semi to crash into us and possibly drive us off the edge of the mountainside. It felt as though we had just gone through a glitch in the matrix. I believe the spirit of the bear was sent to protect us.

Bear can tell us a lot about climate change through story. In "The End of the World" story in the *Book of Dene*, when Bear—who lived in the upper world—stole the sun, he put the sun in a bag and controlled warmth and light. He also had other bags filled with—snow, rain, hail and the like. The animals wanted to release the sun, but the bag of warmth was hung up in a tree on an island surrounded by water and was guarded by the bear and his son. Caribou swam fast to the tree, and the bear chased him in a canoe but he could not catch Caribou because his paddle broke in half. So it was that Caribou got hold of the bag of sun and the animals returned to lower earth with the sun in the bag. It was very heavy and it was a long way from the upper world, so they often stopped to rest. One night, while Mouse was repairing her shoes she accidentally cut a small piece of skin from the bag and the sunlight flooded out, it melted the snow so fast that the earth flooded. That is the end of one story and the start of a new story—when the water animals had to dive down to bring the earth back up to the surface to create a new earth on which animals and humans could live.

As the Manager of Wildlife and Environment for the GNWT, Lee Mandeville is in charge of operations and supervises many employees including my son. He has worked hard to get to where he is today and put his life on the line with his job. Lee started out in forestry as a firefighter when he was a teenager just like many young Indigenous people in the north. He then became an officer and in a matter of a decade he worked up the ranks until he was managing the entire North Slave Region. Each day, Lee oversees an average of ten officers. "No day is the same, every day is different."

The winter season is the busiest time of the year for Lee. "We have a herd that's almost gone," he says sadly. He is referring to the Bathurst caribou herd. "When I first started hunting caribou I was eight years old." Back then there were half a million caribou on the land. At last count the Bathurst herd numbered approximately 5,000. But on the bright side there's been an increase in the Bluenose herd, while the woodland caribou are listed as threatened.

"We ate caribou meat my whole life," Lee tells me. Growing up in the north, caribou was the main diet for the Dene people. One day, he complained to his parents that they were feeding him too much caribou meat, but now he looks back and realizes how lucky he was to have had caribou on the table on a regular basis. It's gotten to the point where some of the younger northern Indigenous kids don't even know what caribou tastes like, and that's a real shame, I chime in. A decade ago, Lee's family stopped hunting caribou. Now they only hunt small game and moose because they don't want to contribute to the population decline. "We became really good moose hunters." His father George once shot a bull when he was by himself. It was getting dark out, but he had to stay with the animal and get the job done. It took him less than an hour to butcher the large animal, quarter it and load it into the boat without any help.

There is quite a serious matter with caribou in the north, says Dene lawyer Jennifer Duncan, who has worked with Elders and harvesters on drafting and implementing a caribou law for her community of Colville Lake in Sahtú. It's something we talk about on our same visit in the city as we make our way along the boardwalk next to the ocean, my dog stopping to sniff at every other tree. He wants to run free, roll in the grass and chase squirrels up trees. Jennifer brings both her legal training and her Dene knowledge into all that she does, and part of her job involves ensuring that the caribou laws in her community are respected by the government. She has done the difficult work of explaining Dene Laws in such a way that the common law can codify them.

Jennifer started the Regulatory Hearing that inspired the the Sahtú Renewable Resources Board to put together a book called *Belare wílé Gots'ę̨ ʔekwę̨.* In English this means "Caribou for All Time." In the book, Elder Charlie Neyelle tells of how the caribou are to be treated, which raises to the larger problem of how rampant agriculture and the taking up of lands are taking a toll on the environment and the migration of the caribou.

In the book Elder Charlie Neyelle tells the story of a Dene couple who had a baby. The baby would cry and cry. The baby cried so much the parents became exhausted. They finally fell asleep because they were so tired. When they woke up in the morning, the baby was gone. They could see his tracks in the snow, so they followed his trail. The baby's footprints turned into the footprints of a ʔekwe, which means caribou. They tracked the prints across the lake to see that their child had joined the other ʔekwe. The parents then understood why the baby had been crying. He'd wanted to join the herd.

There are other versions of the story as well. J.B. Rabesca, Alec Rabesca's dad, the man who survived as a child with his mother Marie alone on the land, tells a version to an interviewer in archival footage while he was alive, in which he explains that a child was crying on the land and everyone could hear the baby, so they went to find the child.

The story as told by J.B. is that no one could find the caribou boy except for an Elderly woman in the community. When the child was taken in by the people, he would cry in the arms of anyone except the woman, so she raised him. The story is starting to sound familiar, and it crosses my mind that maybe J.B. is the caribou boy himself.

J.B. goes on to say that one day as the boy and the woman were fishing the boy got angry and ran away from her, but when he came back he told her that he had killed a caribou. It turns out he had killed many caribou and his hands were bloody because he had put his hand in their throats to kill them. He asked her to make dry meat and caribou fat to share with others. Then he left. The people did not know who had brought the caribou meat but they suspected it was the young boy. They followed the tracks but could only see young caribou tracks.

Lee was told by his Elders that caribou were given to people. "They came down like stars," he says, but he is careful not to quote directly from the old stories because he is not an Elder yet. Dene Healer, Paul Mackenzie, can attest to this. He says that when in need, the caribou will appear if you ask for help. Put down an offering and tell the land what it is that you need, he says. Many Elders have spoken of how, when people were in dire need of food, a caribou would appear. Sometimes the caribou do not even leave any tracks in the snow or mud, says Paul. It's almost as if they fell from the sky. "We must respect the Dene Laws so that the caribou will not be taken away from us."

I too have heard many times that the caribou are like spirits. Hiding in the shadows and over hillsides. They can travel across the landscape like ghosts, and will only appear if they are respected. Many Elders have said that if the caribou are not respected, they will disappear underground. When discussing caribou in meetings, Elders have been known to save a seat for them—so that they have a voice at the table, too.

Sharing is one of our Dene Laws, we must share with one another—so when one has a successful hunt the meat of that animal is to be shared especially with the Elders, children and those who are suffering. Sharing with those who can't afford to purchase expensive store-bought food, or those who have lost a connection with their culture, or those who are unable to go out on the land and provide for their own families, is the most generous gift one can receive, not only for the recipient but for the giver and the animals who have offered themselves to us.

"The Elders said that the caribou can fold the land," says Elder Fibbie Tatti. "All the animals had gifts of their own. Even the raven. All the raven has to do is think about it and the Elders say it becomes a reality." There is another version of the bear and the sun story where it is raven, not caribou, who is able to release the sun from bears control. My friend and fellow northern Dene author Christal Doherty, a Sahtúot'į̨nę woman from Délı̨nę, wrote a children's book on a different version of the story that she was told as a child. In *How Raven Returned the Sun* Raven became a baby in Bear's daughter's stomach and was born a bear. The baby wanted to play with the sun but Bear did not want to let her. The baby insisted until Bear gave in and the baby stopped crying. Raven, disguised as the baby, threw the ball of sun higher and higher into the air until it was released from the teepee and went back into the sky. When the sun was freed, the baby transformed back into Raven and flew away. Bear was very angry but the people were happy once again to have the sun.

When she was younger, Fibbie always made time to visit her Elders and learn about this and other stories. She remembers one old man in particular who could no longer sit up on his own. She would visit him and he would ask her, "My girl, how are the caribou moving?" At the time Fibbie found it strange that he would ask her that, but now she knows why. The hardest thing for him was not seeing the caribou. "The old man knew already that once the caribou start disappearing the land is going to change, and that's where we're at right now." Fibbie traveled a lot at that time in and out of her remote fly-in community and would have been able to see the herds, the old man knew this.

There is one story that Fibbie remembers being told, a story of a time when Raven took all the caribou and the people were starving. They traveled all over looking for caribou, and finally they pressured Raven to release the herd to them. Fibbie feels that the old man was telling her what was to come. That the caribou would disappear again one day and that is what is happening now.

Fibbie is saddened that, in her lifetime, our relatives, the caribou have decided to go home. When she tells me this, she pauses for a long time before continuing on.

The caribou brought such happiness to the people. After a successful hunt the men would cleanse themselves for three days. The women would stand in the moonlight and sing a

The women would stand in the moonlight and sing a love song to the hunters. "How could I forget that in my lifetime?"

love song to the hunters. "How could I forget that in my lifetime?" Fibbie says. She hasn't forgotten, she tells me, and never will—because the caribou is a part of the Dene identity, and when caribou appear out of nowhere, especially a lone animal, that is the caribou giving itself to you. The animals are a gift. "There's no such thing as a lucky hunter."

At one time, animals spoke and Indigenous people could converse with them. Gilbert Cazon's Elders taught him to observe the animals when spending time out on the land. "The animals will act out what Mother Earth is thinking," he says. Once we figure out what they are saying, we can then speak for them, he adds.

Jennifer Duncan reinforced Gilbert's sentiment as we walked along the boardwalk in the city with the sun slowly setting. We are both so far from home yet still speaking about caribou protection. Jennifer thinks that the animals are telling us something but we are not listening closely enough.

"There's no such thing as a lucky hunter."

The GNWT has spent more money on researching caribou than anywhere else in the world, says Lee. It's a start in an attempt to listen to what the caribou are trying to say. As we are speaking, he puts me on hold to order a tea at the local drive-thru, making sure they don't forget the milk and honey. Evening has set in and I wonder if it's his regular nightcap. When he comes back to the phone he takes a sip of tea before telling me he has found a lot of cold cases when it comes to poaching, but the good news is that incidents of poaching are now on the decline. He believes it's because of the conservation work that's being done.

Many people are of the view that the diamond mines have impacted the caribou herds, causing them to shift migration patterns. Lee agrees, but he also maintains that the diamond mines have made an effort to minimize impacts on the local wildlife, habitat, environment and water—and that the north does need an economy.

When I was pregnant with my son, I had a job counting caribou. I flew in a helicopter over the tundra above the diamond mines. The count was zero. Now, all these years later, my son is doing the same work.

Lee has heard Elders say that the caribou have a cycle—sometimes they go up in numbers, sometimes they go down, and sometimes they amalgamate with another herd. Lee is working on a tagging project with the Tłı̨chǫ where they are putting satellite collars on the caribou to track them. They net them as humanely as possible and attach collars to track their migration. It wasn't welcome by the communities at first, Lee tells me, because some people viewed it as a government control tactic, but it is an interim conservation measure that can be necessary when a drastic issue like food security comes up.

Fending off wolves is a large part of Lee's job. Every year he gets about twenty calls regarding wolf sightings that he has to deal with. With wolves being the caribou's main predator, there's been an ongoing wolf cull incentive program in the Bathurst caribou zone. A trapper can get almost two thousand dollars for a single wolf pelt, says Lee.

I thought we might see wolves or a bear at Jennie and James's cabin on Great Bear Lake when I spent time out there, but instead we spotted a lone caribou within the first hour that we arrived. Jennie and I were sitting on the porch of the cabin, our legs dangling off the edge. We were visiting and laughing loudly like aunties when Jennie stopped and said, "Did you see that?" She'd glimpsed something white in the bushes to our left but it disappeared. She brushed it off as a bird, but a few minutes later to our surprise a caribou walked out of the trees and into the clearing. We stayed very still and tried not to startle it. I even held my breath. It was so close we could hear its hooves dipping in the water as it sipped fresh water from the lake. The caribou slowly trotted in and out of the water. We watched in anticipation for at any moment, we knew it could run back into the trees and disappear with the slightest noise. Jennie turned to give the signal to James, who was in the cabin within earshot. The door was open to let in some fresh air, as the cabin felt stuffy without anyone staying in it since the spring. "Caribou" she whispered and it was all she had to say. He dropped what he was doing and grabbed his gun.

We watched the scene unfold in front of us as he quickly yet quietly snuck up on the caribou. The cabin is in a bay, and the sand and rock-lined beach arches in a crescent until it reaches Grannies Point, which is where Jennie's great-grandma used to sit and fish and there is a tree there that looks like a person fishing when you look out at it from the cabin almost as though the memory of her granny has been preserved by the land. By the time James began to follow the caribou, the animal was already on the move. It made it to Grannies Point before James kneeled down, aimed and took a shot—wounding it, but not enough to kill it instantly. The chase was on. He had to track it so that it wouldn't suffer. We could hear gunshots in the trees as Jennie and I paced. He was gone for at least an hour and when we finally saw him heading back toward the cabin. We could see from afar that his head was down. He was empty-handed and I assumed he had missed. I was concerned that he would be upset with himself for missing the shot and so I busied myself in the cabin trying to pretend like it was no big deal. I could see him from the window, he was walking very slowly, but when he got closer to the cabin he gave Jennie a big smile.

We all grabbed what we needed to prepare the meat and jumped in the boat, and drove the short distance to Grannies point where James tied the boat to a small tree. We walked inland to where the caribou was lying lifeless. As an offering I put down some blueberries I had picked along the way up the hill a short distance from the boat. Jennie put down tobacco, and James got to work preparing the caribou. It was a female. I noticed it had only one horn, a unicorn caribou.

The first thing James did was haul the caribou to an open area where there was a natural bed of soft moss. Then he cut off the head. Jennie reminded me not to step over any of the caribou blood, because as women we are powerful and it is said by the Elders not to ever walk over animal blood, just like how we are not to step over a hunter's gun.

James is an expert hunter, having been taught by his Elders. Having said this, there is an imbalance that exists across the vast northern terrain. We see it in the extreme weather causing unpredictability for the animals and the hunters who rely on them for their livelihood. There is danger in unexpected weather events that can catch even the most seasoned hunters off guard when out on the land if they are not careful. If the winter is especially warm, the ice can't be relied upon for travel, and if it snows too much in one season it causes the ice to become unstable. Thankfully it was good weather that day and there was no precarious ice that James had to walk across.

It took him no more than an hour to skin the caribou and pack up the animal parts. Jennie and I helped where we could and built a fire to keep the bugs away—mostly for Russo, their dog, as he was swarmed, his thick coat of fur covered in pesky mosquitoes. We also kept watch for any bears. We would have liked to keep the hide, but it was a smaller caribou—and the fur is not as good in the fall because it is their molting season. If we were to keep the hide, we could have gifted it to someone like Melaw Nakehk'o Antoine, who could have used it in one of her hide-tanning workshops, but we left the hide on the ground where it would eventually sink back into the fabric of the earth after the ravens scavenged the meat on the bones that we left for them.

"Ravens are a hunter's best friend."

I can hear Lawrence's voice as I walk away from the left-over remnants of the caribou, "Ravens are a hunter's best friend." A good hunter will pay close attention to ravens; they will reveal where the moose or caribou is if you watch them carefully enough, because they will circle, he explains. The raven has an all-seeing eye—and perhaps that's why it is explained in the *Book of Dene* that the raven was once perched high on top of a mountain. He was fat and had lots to eat. His beak was red and around his neck he wore a necklace of animal eyes.

Once the meat was quartered, James carefully fit it into a large blue plastic barrel and carried it on his shoulders back to the boat. Sure enough, the ravens picked apart the remains of the caribou carcass once we had left the area. Back at the cabin, we kept the meat under a tarp near the shoreline because it was getting too late in the evening to begin the hard work of cleaning and butchering it. I worried that an animal might come along but James did a good job of covering it up.

Keeping the meat outside overnight reminded me of a story that Elder George Mandeville had told me about how a weasel had spoiled his food supply when he was staking claims in a place called Windy Bay, about thirty miles from the small town of Hay River. He and his friend had been flown into the remote area on a float plane, and had to walk about ten feet in water up to their waists to get to shore carrying their gear. They estimated that the work to stake the area would take them about a month, so they brought enough food to last. They arranged to have a helicopter pick them up a few weeks later, until then they would be completely on their own, without any contact with the outside world.

George and his friend set up their main camp and a separate area to house their food supply in case of bears, but it wasn't just bears they had to worry about. Their very first night a weasel unexpectedly got into their meat supply, which was stored in cardboard boxes. "It peed on all our meat. It stunk like a skunk." Now they would be stuck there for a month with no food. They didn't have any fishing rods to catch fish, but George brought a .22 rifle with him and snare wire, and because of that they were able to hunt small game like rabbits, grouse and porcupine to keep them alive, which George told me latter of tastes like pork.

The next day, James hung the caribou in a teepee and smoked it all day and all night until it was cured. Then Jennie and I began cleaning up the meat by taking the hair, leaves, twigs and moss off of all the animal parts, and cut them into smaller sections.

It is said that the edible organs of the caribou like the intestines, liver and kidney are not to be eaten by children and young people. Similar to throwing wood in the fire without gloves, it is said that you can become an Elder really quickly, so the organs of the caribou are to be kept away from the youth or they could grow old fast. Children are only supposed to eat the meat and a bit of fat, says Dehcho Elder Mary Jane Cazon. Now that she is older, Mary Jane is able to eat and enjoy the organs of the caribou.

My mom likes the liver of caribou and I like the bone marrow, so in the spirit of sharing I brought those parts home to share once my trip to the cabin came to an end.

The caribou that we harvested that day was a woodland caribou, and up until then I had never seen a living caribou up close. Looking back on that day, I think that when Jennie and I first saw the caribou walk out in front of us, it had to have known that we were there because we were being quite loud. Right when we got to the cabin, James had started up the noisy generator. I believe that the caribou presented itself to us that day to give her life to us. Just like Fibbie says, there's no such thing as a lucky hunter. The caribou was a gift that I feel was from my grandmother because it was on her birthday that the caribou walked out into the open in front of us.

I had traveled to Sahtú asking for a sign or message from Creator and was presented with the answers I needed. The spirit of the caribou was there with us that day and I can only hope that I have listened well enough to be able to tell her story.

While on that same visit to Jennie and James's cabin, we went fishing one evening. I haven't done that much fishing in my lifetime, but I would probably be able to figure out how to fish if my life depended on it. I would hope so anyway.

What you need to know about Great Bear Lake is that it extends into the Arctic Circle. It is like the ocean, and once the wind picks up the waves are like swells. That day we would have been caught in those swells if we left the bay we were sheltered in to go fishing so instead James steered the boat deeper into a narrow part of the bay not too far from the cabin. As the sun began to set, we cast our lines into the dark water and waited in silence. Fishing is a lesson in patience. I looked out every so often to where the clouds met the shoreline, keeping a watchful eye in between the trees to see if I might spot a bear or a musk ox. But dusk settled in and my eyes started to play tricks on me. Large boulders started taking the shape of grizzlies and I had to squint to be sure of what I was looking at. I tried to allow myself to enjoy the moment but my fear kicked into overdrive. What if the boat broke down and we became stranded? Would we be prepared? What if I were out here alone having to fend for myself without the help of Jennie and James? Would I know my way back to Délı̨nę? Anything can happen when out in the bush.

My intrusive and irrational thoughts almost got the best of me until I felt a tug on the line and snapped out of it. I panicked and forgot how to reel it in so James jumped up and helped. The fish wasn't putting up much of a fight, and when I pulled on the end of the line the hook was gone and so was the fish. It had gotten away with the hook still stuck in its gills. After a few more casts we gave up and turned around. The last of sun dropped beyond the horizon as we slowly trolled back to the cabin.

I had all but given up on catching a fish until I suddenly felt another tug on the line. James jumped up and helped me reel it in faster this time, so that we wouldn't lose another one, while Jennie stood ready with the net. It was a good size fifteen-pound trout. Not bad for a beginner.

When we got back to the camp, we cooked up the fresh trout with a side of canned beans and fried potatoes cooked over the open fire. It was so delicious that I ate too fast and swallowed a fish bone. Thankfully it went down with a swish of water and did not get stuck in my throat.

Later that evening in the dim-lit cabin I sat under the orange industrial extension cord that was strung up on the ceiling and pulled through the small opening in the sliding window, connected to the generator outside to charge our devices. I looked at the map of Sahtú that was tacked onto the wall behind me where I sat at the table. I noticed that the lake had five main arms. Five channels leading out into other bodies of water. Then it dawned on me. I had once heard a story about a heart being located in the Sahtú. Straight away I opened my laptop, thanks to the technology that allows us to connect to the internet from virtually anywhere in the world, which is not always useful when you want to truly unplug, I was able to google how many main arteries a heart has just to be sure I remembered the few things I learned in biology. Sure enough the human heart has five vessels. I then googled "heart of Great Bear Lake" just to see what would come up. My own heart was now beating fast out of excitement. I had so many questions. Was it a metaphorical heart? Was it the actual shape of a human heart? How big was it? My internet search brought me to a birdwatching website where a man who had visited the Sahtú was told the story of the water heart by a member of the community and he had taken it upon himself to write about it. But the story of the water heart is sacred to the people of the Sahtú, and not everyone has the right to tell the story especially someone who is not Dene. It is up to the people of the Sahtú to decide whether or not they want to share the story of the water heart. It is safe-guarded by people like Elder Fibbie Tatti—who recounts the stories her Elders have shared with her about the water heart in her master's thesis *The Wind Waits For No One: Nı̨hts'ı Dene Asį́ Henáoréhʔı̨́le Ǫt'e: Spirituality in a Sahtúgot'ı̨nę Perspective.*

A version of the story is also found in *"The Water Heart": A Management Plan for Great Bear Lake and Its Watershed.* Long story short, the report explains that spiritual teachers are often mystically tied to different parts of the environment. There was one such teacher from the Sahtú who was connected to the loche, a type of fish that swims like an eel. "One day, after setting four hooks, he found one of them missing. This disturbed him—in those days hooks were rare and very valuable—and that night he traveled in his dreams with the loche in search of the fish that had taken his hook. As he traveled through the center of [Great Bear Lake], he became aware of a great power in the lake—the heart of the lake or the 'water heart.'" He then became aware that the water heart is connected to everything on earth and helps sustain the entire watershed of the Sahtú and beyond.

A heart acts as a pump to send blood through the body and provide oxygen. The heart sends blood to the cells. Once the heart stops, the body shuts down. If water is the lifeblood of Mother Earth than a blockage in the arteries can lead to heart failure on a massive scale when blood is not able to make its way back to the heart. Fallen trees or large dams in riverways can cause these blockages preventing water from being able to flow to and from its main source.

With the NWT being the place where two of the largest

freshwater lakes in the world sit side by side connected underground through a groundwater system, that would make the NWT the record holder for having the largest water body in the world and where else could Mother Earth's heart be located if not for in the middle of what will one day be the last lake in the world according to prophet Ayah. Ɂehtsə́o Erə́ya, otherwise known as Ayah, was born in 1857 and is known as a great spiritual prophet across the NWT. He came into this world with raven power, says George Blondin in *Yamoria the Lawmaker*. He was raised in Behchokǫ̀ but eventually built a home in Délı̨nę, where he lived until he died. Ayah was gifted with the vision of the future. He also had the ability to read thoughts, teach the purpose of life and knew about life after death.

There was a time when we could converse with fish and Elder Jonas Noel knows some of the old stories about talking fish. Jonas lives in the old folks' home in Ndilǫ. I needed a translator to speak to him so I asked my Elderly friend and translator Berna Martin to come with me.

There once were very big fish
and those fish used to talk.

On the day of our visit she meets me on the ramp outside Jonas's unit. She knows him well and introduces us. When we walk into Elder Jonas Noel's living room, the first thing I notice is that his walls are covered in photos of him and his family out on the land hunting, and the corners of the picture frames still have cardboard on them. I look around at the many framed photos on the wall and see that there are none of him fishing. He must be more of a hunter than a fisherman, I think. Or maybe he is related to the fish. Jonas doesn't waste any time and seamlessly switches from one story to another as Berna helps translate his fast spoken words into English for me. His father passed on stories to him that the Elders would have told his father when he was young and one of those stories was that there once were very big fish and those fish used to talk. Jonas tells me that not only did fish talk, there was also a time when they gave people their names. At that time, little bear cubs talked too. He adds with a chuckle.

He speaks for a very long time without taking a moment to pause, giving Berna a run for her money having to remember everything he is saying. Once there was a mother bear who told her two cubs to stay put while she went to check for berries. While she was gone a fish came around and killed her cubs. This was at a time when fish walked on land. When the mother came back she was very upset to see her cubs were gone. She had something heavy in her hand. When she found the fish, the fish was sleeping. The bear said to the fish, "Where should I start to kill you?" and the fish said, "Start at my head," but his head was made of steel. The fish and the bear started fighting until a big tree broke and both the fish and the bear fell into the water with the tree. The mother bear said to the fish, "From this day on you will live alone in the water and you will never have anyone with you."

The fish may have been punished with having to live alone in the water forever, but they may be making their way back on dry land. During a phone call with Gilbert Cazon he told me that, about thirty years ago, there was a man checking his fish nets in a place called Shale Creek, north of Łíídlįį Kų́ę́ about sixty kilometers down the river. He would check his fish nets early in the morning, and one day he noticed strange tracks he had never seen before coming out of the water to the edge of the shoreline. "He would see these tracks every once in a while." The man asked some Elders, and they said they'd never seen tracks like that before either, but one Elder told him that it was getting to the point where animals were going to start changing. "When the water gets bad then the fish grow legs and feet and are able to walk out of the water and eat proper vegetation from the edge of the shoreline." The man eventually caught a fish near that same spot but it wasn't just any old fish. It had feet—eight, to be exact. He gave that fish to the local government office and they sent it out for investigation but the man never heard back from them about what type of species it was.

With animals adapting to the warming environment, animals not normally seen in the north are now making the colder climate their home. Magpies, coyotes—even a dove was reported just west of Nahanni Butte. In fact, almost everyone I've spoken to in the making of the book were concerned that certain animals are migrating further north. Elder George Mandeville has spotted hummingbirds, and sadly he doesn't think they'll survive the winters. And if we don't already have enough big-game predators, we've added one more to the list: the cougar.

Nearly a decade ago, Steven Nitah was snowmobiling in the late spring. The snow was melting and there were areas of open water on the lake. He was about ten kilometers outside of Łutsël K'é, going through a portage in the back bay toward the place that he was born. He stopped when he saw what he thought was a lynx bathing in the sun. The animal looked right at Steven before getting up and slowly walking away. That's when Steven noticed the long tail. He realized it couldn't have been a lynx, because lynx have short tails. Steven went back to Łutsël K'é to grab his phone, then went back and took photos of the animal's prints in the snow only to see, to his surprise, that the big cat had followed him. He compared the print with a lynx paw and a cougar's paw, and there was no mistaking that it was a cougar. It was not the first time that Steven had heard about cougar sightings that far north.

With cougars making their way into the northern terrain, my great-uncle Alfred Baillargeon, Randy's grandfather, has told me that there is a big lion's tail in the earth at the opening of the Tibbitt to Contwoyto winter road, which leads directly to the diamond mines out in the tundra. The landmark is from long ago says Alfred, and I wonder if it is a prediction that one day cougars will travel so far north that they will make their way to the divide in the tree line where the boreal forest meets the tundra—or maybe they were already there once before long ago, when the north was warmer, and now they are making their way back.

Alfred told me that before the two rocks at Tibbitt Lake were blasted in order to make a road to the diamond mine, there was a giant lion lying in that place for many years. The lion was inside the rock. Did the explosion cause the lion to be released after all the diamond mines opened up a few decades ago? This would line up with around the same time that cougars started being reported in the north.

There have been quite a few cougar sightings in Echaot'l Koe because the community is closer to the border of British Columbia, where cougars naturally roam. But cougars don't care about borders; they are merely following the deer, who are also making their way further north.

With the deer population on the rise, we have more ticks coming up north now too as they catch a ride on the deer. Some ticks are surviving through the warmer winters in the north, whereas before they would not be able to survive the cold. If there's anyone who can tell us about the prevalence of ticks in the north it is those who work closely with animals. People like Melaw Nakehk'o Antoine, hide tanner extraordinaire, who says that the process of hide tanning itself is very sustainable. "We do everything as traditionally and naturally as possible." Hide tanning is an Indigenous climate solution, explains Melaw, because hide is a very resourceful material that is not only used to create artwork but also supports Indigenous economic systems like trading and making clothing for families and community members.

The process of hide tanning varies across the north. Some use the brains of the caribou to soak the hide to give it a smoother texture. Some use the fur to create blankets or do tufting work. As a hide tanner and instructor Melaw has traveled all across Canada to many Indigenous communities, creating, knowledge-sharing and developing lasting relationships. The process of tanning hides brings the community together, and for Melaw that is what matters. Hide tanning for her is an act of resistance.

Melaw, who, among her many talents, starred alongside Leonardo DiCaprio in the movie *The Revenant*, says, "We're using the land, we're working closely with the animals and we have to go out into our traditional areas to harvest and hunt." Through the actions of being on the land and observing, "we are intimately and consistently on our land and seeing the changes that are happening."

Melaw sees herself on the front lines—witnessing both the slow and rapid changes. "Because we work so closely with the skins of animals, you can see the determinants of health." There are some animals that have ticks on them, which Melaw says is unusual because the winters are still cold and would cause the ticks to die but, in some cases, they are wintering on the animals and surviving.

Through her work, Melaw has learned of a brain disease in deer called "chronic wasting disease." The animals are stressed—and the more that the animals are stressed the more susceptible they are to infectious diseases. There are also more cases of moose rubbing the hair off their shoulders, she says, and this means there are more pests bothering them in the summer. "There are different types of indicators that you see on the skin of the animals if you've worked with them and know them intimately." These are details that you wouldn't see otherwise.

We live in a society of complicity, says Melaw. We have a short attention span, and we disregard what is happening to the earth—or we do not take seriously, or feel like there's nothing we can do to stop it. A dream of hers is to one day get enough funding to track the incidences of disease that she comes across in her work of tanning hides, and to have a scientific paper written about the findings—as well as the climate impacts on the environment and the animals that she works with so closely.

A few years ago, Melaw broke her hand and had to take a break from hide tanning for a year and a half. Working with hides is no easy endeavor. It is hard work. Sometimes her hand still aches if she works too hard. "I guess I still have to strengthen it. For a really long time I was doing everything with my left hand so I did a lot of left-handed paintings."

During the Covid lockdown, Melaw and her family lived out on the land. When I ask if she would go back out and live on the land again in the face of an existential crisis, she answers that she probably would and asks me if I have read Waubgeshig Rice's postapocalyptic novels. I nod and she adds, "This is why we have to continue on with our teachings." And I couldn't agree more.

Melaw was making moccasins with a group of high school students once, and they asked her that same question, "What if something happens to society?" And Melaw answered: "Well, you're all going to know how to make moccasins to take care of everyone's feet."

Gilbert Cazon tells me that his Elders would say that the river will one day bring animals that will save the land for us humans—and that is what is happening now. Gilbert says that rather than thinking animals on the move are invasive, they are actually supporting and protecting the areas that need it most. This is an entirely new way of looking at the term invasive species. The Elders would say that when animals are going through a period of devastation, other animals will appear in habitats that are not natural to them—and they will reestablish a balance in the ecosystem. It's just like humans, says Gilbert. "When we build a house we have a carpenter, we have a journeyman, we have a plumber, we have an electrician, we have all of them doing one thing to build that house."

A healthy environment runs like clockwork. "The first thing that happens is the rain comes. It softens the land. Then the sun comes and it hardens the land. Then the animals with sharp claws like wolves will run all over and open up the hard ground with their claws." Then the rains will come again and be absorbed into the surface, and "once that happens the larger animals that have hooves will come and walk all over the land making divots in the ground, then the winds will blow the seeds into the divots, and once the seeds grow, we the Dene come along behind the animals following them and pat down as we walk around the area until regeneration happens."

Certain animals that we deem as invasive may have once already lived in Dënéndeh and their return is needed, says Gilbert. We should allow them to do what they need to do and not look at them negatively. The specialists coming to our land do not know the land the way the Indigenous peoples of Dënéndeh know the land, because they are only here for a short time he says. We should not think we can manage the land; the animals do that for us. The specialists that come to Dënéndeh for a little while and leave again do not know our teachings and they should not be trying to play God.

What I'm hearing Gilbert say is that trying to manage animals could throw off Mother Earth's natural maintenance of the land, and we should not interfere with that. Those who are trying to tackle invasive species might think they are doing a good job, but they could be making things worse. Gilbert uses the example of bringing buffalo from the southern plains to the north, which initially caused an imbalance in nature and the spread of diseases like anthrax, which led to massive extinctions. Physical changes in animals are markers that tell us the land is changing, and so do behaviors such as migration patterns, says Mary Jane. There doesn't seem to be as many types of birds and waterfowl as there once were, but the geese still fly north in the summer which is a good sign.

The world would never be the same if we had to grieve birds entirely or any animal for that matter but Chief Fred says that one of the Dene prophets from long ago warned that "if the birds are silent and they disappear, something's going to happen." And that is what Fred is watching for. "That day of silence." Animals can sense danger and the birds are the first to tell us if something is going to happen—so watch the birds, he says. "That day of silence, one day will come."

Come to think of it, I have not heard the harmonious sound of the white throated sparrow, or the Ts'ihts'ats'a—which translates to "sparrow" in Dene Zhatıé—in a very long time. The beautiful but formidable magpies who are arriving in droves from warmer climates are the main cause of their decreasing population.

The sound of the morning birds might be an annoyance to those who want to sleep in on a Sunday morning, but to others they are a gentle alarm clock welcoming a new day. Elder Jonas Noel, in his small unit in the old folks' home across from Chief Fred's house, laments about the lost sounds of the beautiful birds singing in the spring and it is not attributed to his hearing loss.

We truly don't know what we have until it's gone. Whenever I walk past a tree full of noisy birds chirping at one another as though they are having an important meeting about the state of the world I stop and take a moment to take it all in, to listen closely to what they are saying before they fly away.

— 5 —

Mother Earth Is Our Elder

For me, the north is like a wise Elder that I visit time and time again. I too will one day be an Elder if I'm meant to live that long, and I plan on living the last days of my life out on the land in the north growing old alongside Mother Earth. When our lives end our bodies go back to the land, like a house that's been abandoned, Mother Earth slowly starts reclaiming our physical bodies. We witness this in the aging process of our parents and grandparents to prepare for the inevitable.

When my grandma passed on into the spirit world, I regretted not spending more time with her. She was so full of knowledge and now I am left to piece that wisdom together, but it doesn't have to be this way—we must make sure that there is a knowledge transfer before our Elders pass on. It is very important to listen to our Elders, it is one of the Dene Laws.

My grandma spent her entire life taking care of others. Being the eldest sibling of sixteen children, she had the main responsibility of helping her mother raise her brothers and sisters. She then went on to have six children of her own. Then, when she thought her job of raising children was done, she took me and my sister in and raised us too, saving us from being lost in the foster care system.

Even in her later years my grandma was as tough as a buffalo, that was up until she had a bad fall and hurt her hip outside of the post office. After her fall she began to experience such tremendous pain from the nerve damage that she had to undergo back surgery. I remember how afraid she was. I tried to console her. She had never had surgery in her life and was so worried about something going wrong that she would rather continue to live in pain. As strong as she was, I could see that the pain was getting the best of her and she had a very low quality of life. I tried to reassure her that she could trust the doctors and that they knew what they were doing and after some persuading she finally agreed, and I was there with her every step of the way to comfort her in the hospital.

Once she recovered, she was herself again. She had no pain for many years but then she had another bad fall and on top of that she had her house broken into when she was alone one night and had to fend off a burglar. She was almost eighty. After that she started to sleep a lot more and wasn't eating very much. I tried to visit her when I could, but I was so busy with my own life that I often didn't make the time. I would sometimes go visit her when my kids were sleeping and lie down beside her and help her use the washroom. We couldn't afford home care to come in and take care of her, and our family had trouble coordinating shifts. We weren't able to be there for her in the way that she needed. We failed her. I deeply regret the times I saw her phone number pop up on my phone but didn't answer because I was "too busy."

Then one day I went to visit her and saw just how frail she was. I could not watch her wither away in front of my eyes and do nothing. I brought her to the hospital where she was given a room in the emergency ward. As much as I wanted to stay, I had to leave her and promised to come back later. She was in good hands, I thought. I brought her a blanket so she didn't have to use the hospital blankets. I waved at her from the hallway. It was the last time I would ever be able to speak with her again.

Her departure out of this world and into the next should have been without suffering. She should have had her entire family by her side. Often times when a Dene person is not well and on their death bed, the entire community comes to the hospital. When my great-aunt was dying of cancer there were so many people inside her hospital room and outside in the hallway it was standing room only, but there was no one there for my grandmother until it was too late. Isn't this how we are also treating Mother Earth? She is struggling in pain before our very eyes yet we are not there for her. We are continuing to turn a blind eye to her pain when what we should be doing in her time of need, is to be there for her.

Looking back, I ask myself why I was so absent minded. At the time I didn't think of what life would be like without my grandma in it. And when she died it was too late to go back and make up for all the lost time. I wasn't there for her when she needed me the most.

I still hang onto that regret to this day. My grandma was my mother; she was my whole world. Just like we grieve for our loved ones, so too will we grieve for Mother Earth. We will grieve the water, the grass, the animals, the wind, the rain;

we will grieve the time we had with her, we will grieve the beauty we take for granted. When we begin to view Mother Earth as a respected Elder and loved one who deserves reciprocity, we can begin to take responsibility to care for our loved one and share in her suffering. We must not allow ourselves to get so caught up in our own lives that we forget to notice that Mother Earth needs us.

When we have children, we expect that they will be there for us when we are older. We care for our children when they are unable to care for themselves, we help them tie their shoelaces, we catch them when they fall and in turn they are supposed to be there to care for us when we are no longer able to do things for ourselves. Nowadays there are reports of Elder abuse in communities with social isolation on the rise, some family members are taking advantage of their Elders by taking their money. Similarly, we abuse Mother Earth when we take from her. We must check in on our Elders from time to time to so that they do not become lonely. We must bring our children to visit with the Elders and sit down quietly to listen to what the Elders have to say. We must do this for Mother Earth too. We must bring our children to the old growth trees and the ancient rocks and sit and listen to what Mother Earth is telling us. Children and Elders are furthest apart on the spectrum of life, but they are closest to the spirit world and that is why the bond they share is sacred. It is a bond that has been interrupted for many people in today's society, and it stems from the nine-to-five working world we have found ourselves in. Prior to first contact, Dene people worked out on the land. Everything they did, they did together as a family unit. If one or both parents set out to hunt, fish or trap, someone from the extended family would step in and take care of the children—usually a grandparent. There were no strangers being paid to look after their children in a daycare all day while the parents went and worked in an office.

We have been so blinded by the belief that we must work to make money that we forget that our lives have meaning—and most often our purpose is related to a higher calling for the benefit of all, and not for the accumulation of wealth. The main purpose of life is to be of service. To care for each other.

Now everyone has to work hard to make money to afford to live and even sometimes that's not enough for the average person to keep a roof over their head. Many of the jobs that take people away from their family are trade related. Our Elders have witnessed the industrial revolution from the very start, which has resulted in the encroachment of their lands and the pollution of the environment in the name of progress. Capitalizing on Mother Earth's resources—namely fossil fuels, which are really just plants and dinosaur bones when it comes down to it—is irreparable, and no amount of money can fix the damage done that has resulted from commodifying nature. Ask any Elder and they will tell you that the health of the land and water is priceless. That in fact the word *progress* is not about progressing in the way that we think it is. It's about taking more and more from the land until there is nothing left to give. "Progress" is not moving humanity toward a better future, but moving us away from the ability to live in harmony with Mother Earth—and it's making us sick.

We are causing our own extinction by continuing to operate in a way that is not in alignment with the balance of the ecosystem, and Dene Elders can see this more definitively than anyone else because "progress" has incrementally, over a lifetime, impacted their ability to fully live out their cultural practices and belief systems. Dene Elders document this change over time orally when they tell us about their childhood, which was once rich with uninterrupted access to fresh water, fresh air, fresh soil and healthy animals. They lived in relationship with nature which helped keep the ecosystem in balance. As children, our Elders were able to cup their hands in any stream of moving water and drink from it for free. Today, most water bodies have been compromised. Nowadays, we would never think to drink from a body of water without wondering if it's polluted. We have become thirsty for the way things were, and we don't even know it because we have normalized the sad reality that most everything in our natural environment is now tainted. Dene Elders remind us that our new normal is in fact not normal at all, and is something we should not become accustomed to or complacent of.

I have many Elders I admire and look up to. Elders like my great-uncle Alfred, who is my grandmother's half-brother. I visited him one New Year's Eve in his home up on a hill in T'èʔehdaà while a drum dance was taking place at the community hall just a few hundred feet away. Once again my Elderly friend Berna Martin, who is Alfred's neighbor, accompanied me so that he did not have to resort to speaking in English for my sake.

Alfred tells me that the winter weather is milder than ever before. It is expected to be in the −40 to −50 range in January. That night it was around −20. He says the warmer weather

> What humans have created with their own hands is destroying the earth and that is what is causing global warming.

from down south is coming. Past Elders predicted that the warm weather will cause sickness. That the fog that comes up from the ground and through the wind carries and traps disease in the air. It makes me think of the thawed animal corpses that have made the news in other colder climates that are also experiencing the extreme effects of global warming—like the discovery of an ancient reindeer carcass in the thawing permafrost in Russia, where a family was exposed to anthrax spores and a child died as a result.

There's a lot of damage caused by non-Indigenous people says Alfred, and the atmosphere has changed because of it. Creator made the earth, and humans made it possible to leave it by going to the moon. What humans have created with their own hands is destroying the earth and that is what is causing global warming.

He points out his window. The sunset used to be over there, he says, but it has moved. Alfred has stood in the same place all of his life and watched the sunset; he of all people would know if the sun was setting in a different location on the horizon as he looks out at a perfect view of Tinde'e. He has heard professors talk about the atmosphere, the sun, the ocean. Alfred is no scientist and may not be able to logically explain how, over the course of his lifetime, the sun has changed position, but he is right. With the melting of the polar ice caps, the earth is rotating differently due to the liquid core of the earth tilting it on its axis, thus changing the location of where the sun normally rises and sets. I think back to when I was a little girl when I would spin myself in circles so fast to try and catch up with the earth's rotation but it just made me fall down with dizziness.

The world now is literally off balance. The takeaway? Always listen to your Elders. Alfred has been saying this for years. Scientists are only catching up to what Dene Elders know through observation.

Elder Jonas Noel grew up in Enodah, the place of many lynx. He used to travel with his dad all over Dënéndeh to trap, fish and hunt. He and his dad chopped down logs at Nı̨hshìì and hauled them with a small boat and kicker to Enodah, where they built themselves a house to live in. In the spring Jonas set muskrat traps with his mother. He got a lot of furs back in the day, he says with a smile. He tells me that one time he and his mom caught 400 muskrats in a day. They had an old teepee made of canvas material that they made a packsack out of, and put the muskrats in to haul them back to Enodah on foot. They were resourceful. At that time they had to go far to trade the fur pelts for rations.

People worked very hard back then, Jonas says. It was colder in those days, he adds. He would have to chop the ice with a chisel six feet down just to reach the water to put a net in—that's how thick the ice was. They caught lots of healthy white fish and trout. He opens his arms wide to show me how big the fish were but not near as big as the talking fish he describes in his story. The Dene ate good back then and so did their dogs. People traveled quite a way for sustenance. Jonas reminisces about the last caribou hunt he ever went on with Alfred. They had dog teams back then he says that could beat anything with a motor. They shot a lot of caribou that winter. The caribou were just about to head back into the barren lands and they got them just in time.

Jonas speaks of a time in the future when the lakes will no longer freeze over. Jonas's message for the world is that the atmosphere has changed and the sun is getting hotter. The earth is becoming hot like the sun. His father used to tell him that the sun was coming down to earth. Jonas says that the young generations need to be taught to live on the land—it's the only way they will survive in the years to come. But above everything, only Creator knows what will happen in the future.

When I meet up with Paul Mackenzie in the busy restaurant in the middle of winter in Yellowknife with my daughter, mom and Berna, we drink so much tea among the five of us that the server brought out a large silver kettle full of hot water and tea bags so that he wouldn't have to keep coming to our table every few minutes to top us up.

Paul is the same age as my mother. Even though I consider him an Elder, he says he is not quite an Elder yet. He is, however, a well-known, respected healer in the community, and forages in the bush harvesting natural plant medicine to treat various illnesses.

When Paul was a teenager, an Elder told him that our ancestors were alive when the whole world flooded, then froze over, then melted, and to this day the water is still receding. There is undisputed evidence of a great flood, and scientific evidence that after that there was an ice age. In just a few simple words while he seasoned his mashed potatoes with salt and pepper Paul explained what occurred on our planet over the span of millions of years. It did not take him obtaining an expensive degree or studying the earth's core, its gravitational pull, or the layers of sediment in the ground to come to this conclusion; it took the passing down of Indigenous knowledge systems through oral history and the protection of that knowledge over millennia through story because they were there when it happened.

The way Elders have explained global warming to Paul is that there will be bad weather, and everyone will get sick from it. As he is telling us this, Paul is interrupted when his phone rings. He ignores it at first but takes the call after a few more rings. It's a flip phone, and that isn't what surprises me. What surprises me is that he has a phone at all. When he hangs up and addresses us again, I have to strain to hear him over the clanking dishes and people talking. He speaks mostly in broken English, which gives Berna a bit of a break from having to translate.

He tells us about a dream he had about a volcano. "I was

standing right at the volcano," he says. The Precambrian shield that extends across the North American continent is partly the result of volcanic activity. The NWT has regions of molten rock where volcanoes produced kimberlite magma that formed those rocks. There could be an active underwater volcano under one of the great lakes in the NWT right now and we wouldn't even know it.

Tucho is deep, measured at more than 2,000 feet. Some of the advanced methods that submarine volcanologists use include human or robotically controlled submersibles and satellite observations. While these technologies may be able to see the bottom, to this day the lakebed of Tucho has not been fully explored for the simple fact that the area is too expansive and it would be too expensive to cover. Paul says that if the water continues to recede it will expose the underwater volcano that he dreamt of and cause it to erupt. In the worsening drought that the north is experiencing I wouldn't be surprised if a volcano eruption did take place. Could it be that we are living next to a ticking time bomb in the NWT?

Right now the water in the back bay of Tucho is the lowest it's ever been on record. So low that houseboaters will soon have to pay property taxes if their boats are no longer floating and instead become beached on dry land.

This is all to say that we must pay attention to what our Elders are telling us, as they have lived through decades of change and carry knowledge more valuable and accurate than any researcher or scientist. They have quietly stood by and watched as the water levels have receded over the years and they have tried to warn us but once again we didn't listen or do anything about it.

Environmentalists and researchers who visit the NWT for a short time will only leave with fragments of the truth if they do not include the knowledge of the Elders. Our Elders have carried oral histories with them their entire lives. Just because they are not written does not make them less important, or inaccurate.

My cousin Tommy says that scientists will claim that if it's not written, there's no record, but the Dene didn't have written records—the minds of our Elders are where the records are kept. "Every Elder is a library," he exclaims. Tommy believes in science but he also believes in the stories he has been told by his Elders, and when it comes down to it, "Science itself is still pretty young… so I'll take an Elder's knowledge over science any day."

"Every Elder is a library."

Our Elders are the historians. Our history is so well preserved that it is to be protected for the sake of our way of life. In the busy restaurant, Paul tells me that his mother used to talk about dinosaurs. When she was young her and her dad saw dinosaur ribs in the Gamètì or Rae Lake area. Her dad told her not to tell anyone and she didn't for a long time. She only told Paul about it in her later years. I'm guessing that Paul's grandfather didn't want anyone to know there were dinosaurs in their community because it would have brought many people from all over the world and they would have to dig up their land and disrupt their peace.

As we were driving home from the restaurant, I asked my daughter what she thought about the conversation we had with Paul over dinner—and more importantly what she thought about being in the presence of a healer. Like a typical teenager she shrugged and said, "He's cool," and looked back down at her phone.

But dinner with Paul was like meeting the Oracle in the movie *The Matrix*. It might have seemed anticlimactic, but sometimes the greatest lessons are so subtle that we seldom notice how impactful they are until we look back on it. Even though it might have seemed like Paul's words had fallen on deaf ears, I believe that somewhere in her subconscious my daughter internalized his words—and the words of many Dene Elders she has had the opportunity to listen to and learn from along the way, during the creation of this book and in her years growing up in the north. This is an example of how important it is for children to sit and listen to their Elders. Even if it may not seem like they are listening, they are absorbing the knowledge that the Elders are sharing—and the more they are exposed to sitting down and listening to their Elders, the better they will be able to understand the world and their place in it.

Back in Łutsël K'é at Elder Celine Marlowe's house, she is listening to the radio and looking up from her sewing from time to time to glance out her living room window. This is her routine most every day—that is, if she's not out at her cabin. From her house she can see everything that goes on in the center of the community. In front of her on the kitchen table, is an array of colorful beads and pieces of hide spread out over a dish towel.

We watch as the government worker walks around the community looking lost. He was scheduled to fly to the community on the same charter as me the day before, but missed his flight. A tractor barrels down the road behind him towing the backwards water truck and the government worker jumps out of the way. We laugh. I wanted to ask Celine why the water truck is being towed backwards, but she starts talking about climate change, knowing that is the reason I am there.

"No one is doing anything about climate change," she says. I can't help but notice that the walls around her are painted a bright green. "I remember as a child there was lots of snow in April. We used to slide on the hard surface of the snow." Now in spring the snow is melting around the trees. It never used to be this way. "It's thawing way too early." It used to be that there would be no puddles until the summertime, she recalls.

She looks over at my muddy shoes by the door, and then at my jeans which are noticeably dirty from the litter of puppies that I couldn't help but stop and pet on the way over. Clearly, I met a few puddles on my short walk to her house from the blue house on the hill and I'm slightly embarrassed at my appearance. "The weather is not even cold. All winter it's mild," says Celine. In the olden days, it used to be in the −60s, windy and cold. The sound of Dene drummers blares from the little radio beside her on the table.

Celine was just out at her cabin a few weeks ago, and she tells me all about the caribou hides she has hanging outside to dry.

Celine was born out on the land. Like many Elders, she lived without running water and electricity. Most people use fuel now to heat their homes in the north, but she would much rather have a wood stove. She worries about the pipes freezing because she no longer has a woodstove. In fact, she was promised one two years ago by housing, and is still waiting for it to be installed. When she returned from her cabin a few weeks ago she was indoors sewing away and noticed that her hands and feet were cold and she couldn't warm them up. She and her grandson went and took a look at the furnace and realized that only cold air was coming out. Normally she would be able to fix the furnace herself by bleeding the lines if it kicked out, but because she had been given a new furnace, she wasn't able to figure out how to start it back up.

"Folsom Prison Blues" starts playing out of the small radio. When I ask her how much time she spends at her cabin she quickly says, "As much as I can. Tomorrow I'm going." Celine is planning on checking her muskrat traps, but says she doesn't trust the lake ice anymore in spring so she just stays in the bay close to the shore. One year she put a trap out not too far from the shore because she knew there was a muskrat there. She went to check the trap and the muskrat was still alive with its paw caught in the trap. She didn't want to grab it with her bare hands because they bite, so she pushed it back in its hole and unfortunately its paw was dismembered. Trappers never want their animals to suffer; they hope for a quick and painless death. Two days later she caught the same muskrat and noticed that its wound had healed. She tells me she's lucky if she gets six or seven muskrats when she sets her traps. She smokes them and gives them away to people.

Being out at the cabin and being active in her cultural lifestyle means that Celine can maintain her physical strength as she ages. Even though I'm half her age, I still have a hard time keeping up with her. Not only is she fit, she is fearless. I was fortunate enough to be invited out to her cabin one evening. I had rented a four-wheeler to get around the community from the local Co-op and met her outside her house. She was already ready and had been waiting for me on her idling four-wheeler raring to go. I was a bit afraid to venture too far out of town not knowing which way I was going so I followed closely behind Celine for fear of getting lost in the bush but I might have followed a bit too closely because the dirt that was kicked up from her four-wheeler got in my hair, mouth and eyes. By the time I got back I looked like I was an Elder myself because the dirt filled every line in my face to the point that one might think I was carved out of wood. It's no wonder that most people drive around on their four-wheelers with scarves over their mouth and nose and wear sunglasses. I was thankful Celine had a rifle tucked at the ready next to her foot. "Have you ever seen a musk ox before?" She yelled over the sound of the sputtering engines. I shook my head. Not in the wild anyway. She slowed down at the top of a hill overlooking a meadow scanning the valley below. She must have the eyes of an eagle because I would have missed it completely if she hadn't pointed it out. Standing next to a cluster of bushes was a musk ox. It was so big that I thought it was a grizzly bear at first. We stood there for a while and watched it graze and I was glad we were a safe distance away from it. On the way back from the cabin, as we headed into the blinding sunset, I looked to my right and for a moment I saw a large dark figure in the trees. A part of me wanted to stop to make sure what it was but I already knew. It was a musk ox. I could see that it had horns that looked like a helmet. It had to have been no more than ten feet away from the trail. I didn't stop. Celine was far ahead of me and I was too afraid to turn around to get a better look. I regret that now. I will always remember that moment and wish that I just faced my fears and turned around to see the beautiful animal standing in front of me in all its glory.

You wouldn't know how strong she is to look at her, as she is quite petite, but Celine is very fit from being out on the land nearly every day. Driving the four-wheeler to the cabin and back hurt my hands after a while but for Celine it's how she gets around in the spring, summer and fall so I didn't complain. In the winter she sometimes has to get out to her cabin by skidoo. One year her niece lent her a skidoo to get up to the cabin, because there was too much snow to take her four-wheeler. Celine forgot that the pull rope to start the engine on the skidoo was short and she pulled it way too hard. She brought the skidoo back broken and handed her niece the piece of rope that had ripped off with an apologetic expression. Some people, including myself, have a hard time starting a skidoo with a pull rope, let alone yank it so hard that it snaps in half.

One of the only reasons Celine doesn't live at her cabin full-time is because her family worries about her being out there alone. "I'm so thankful I have a cabin," she says to me as her little dog Scruffy, who is the definition of what I would consider a scruff to be, comes up to her. He is an indoor dog, not like the huskies running around freely outside guarding the band office.

The community of Łíídlı̨ı̨˛Kų́ę́ is on an island in the Dehcho in the shape of a large animal claw from above. Gilbert Cazon says the soil is different there compared to the rest of the NWT, and at one point in history Łíídlı̨ı̨˛ Kų́ę́ was the garden capital of the Dehcho. Born and raised on the land, Gilbert lived on the banks of the river in the summer, and in the winter he lived inland. He fought forest fires for over thirty years. So when it comes to seeing changes on the land he can speak volumes. He used to be able to put out a five-hectare fire in a day that was because he had permafrost on his side. Now, with the permafrost melting at rapid rates, the same size fire would take at least ten days to put out, he says, because the fire is able to travel deeper into the ground.

Once, at a community meeting, the late Charlie Snowshoe Sr. said something that really stuck with Gilbert: "It's not climate change, it's man-made change." The cumulative effects, says Gilbert, are a result of the things that have been done to the land by humans. It's time to listen to the Elders and Indigenous knowledge carriers says Gilbert. "As harvesters we, we know the stories from long ago, we don't need to analyze things, we can predict what is going to happen in the future." It is true, Dene Elders have predicted significant events in history that have already occurred and there are many more prophecies that are yet to come to pass.

"It's not climate change, it's man-made change."

Growing up, the Elders would always say to Gilbert that the world was going to go through a cycle of change. Now an Elder himself, he tells me that climate change is part of that cycle. Humans have sped up this natural cycle with greenhouse gas emissions. Gilbert says that his Elders have informed him that heavy snowfall is cleansing for Mother Earth because it acts as an insulator. I imagine the snow as one big blanket keeping Mother Earth warm for the long cold winter months.

Mary Jane and Gilbert have known each other their whole lives. They were born a month apart. When their mothers were pregnant, they were in the hospital together. At that time pregnant women from the Dehcho communities had to go to Łíídlįį Kų́ę́, where there was a central hospital, and stay until they delivered their babies. Once their babies were delivered they would have to wait for the first boat to come in to take them home. Nowadays women are still sent out from remote communities to give birth, only this time it's by plane and not by boat. There is a movement of Indigenous midwives bringing Dene birthing practices back into the communities and more and more women are now opting to give birth at home.

Because they were the only two Dene women in the hospital at that time, Gilbert and Mary Jane's mothers were segregated in the same room as they waited for the boats to arrive to take them home. According to their mothers, the babies would cry if they were by themselves in their cribs so they put the babies in the same crib and they would no longer cry.

Gilbert and Mary Jane went their separate ways when their mothers returned to their camps at opposite ends of the Dehcho. They didn't meet again until they were seventeen, in 1979. Gilbert was on a break from residential school and Mary Jane had just come out of residential school. Gilbert went to the northern store. He wanted to pick up some items before going out to fight a fire. That is when he saw Mary Jane. It was love at first sight. He was so enamored by her that he asked his friend who she was. His friend didn't want to tell him because he liked her too, so Gilbert worked up the gumption to introduce himself and the rest is history. "When we got married, that's when our mothers stood up and said we need to tell you both something." The first time they met was when they were in a crib together, and their mothers couldn't believe that they'd found one another after all that time. Their special bond gives me hope that soulmates truly do exist.

Being raised on the land, Mary Jane says she had to get up very early in the morning. All the family members used to gather at the river before starting their day and make an offering. If they didn't have tobacco they would make their own offering, mixing fireweed with red willow bark—which was always in abundance. They would feed the water and talk to the water spirit, the sky people, and all the advisors in the supernatural world. They would ask for help for that day.

Once they were done praying, they would go to the shoreline near the tent and gather around and feed the fire, asking their ancestors to help guide them for that day. Then Mary Jane would be given a cup of hot broth. "It could be caribou broth, moose broth, rabbit broth, wild chicken broth, fish broth, duck broth, goose broth, porcupine broth, squirrel broth—whatever broth was available that day, they would give us some of that broth and they always made sure it was warm." She says even now she drinks a lot of broth.

Broth provided the energy needed to be out on the land. Broth is full of nutrients. It's what marathon runners drink to keep hydrated during long races. Mary Jane's Elders would say that if you ingest cold liquid you can get sick. You have to keep liquid at the same temperature as your body temperature.

At about six in the evening, after their ceremonial gathering, Mary Jane and her family would all go out on the land and check the rabbit snares. If a moose or caribou was shot that day, they would camp out to make sure all the meat was properly prepared and then they would haul the meat back to the main camp. The younger children would only pack a piece of liver or a kidney or a couple of ribs. They would only take as much as they could carry. As she was growing up, Mary Jane was taught when to harvest and how to prepare dry fish, dry meat, pemmican and other harvested foods. Juniper, for instance, would be dried then pounded down into a fine powder that was used to season food like pepper because it has a similar kick. It's important to remember that this was at a time before there were any processed foods available to her family, and no grocery stores in sight.

Everything that Mary Jane was taught was to be done with purpose and care just like her sewing. When preparing hides, everything that would be used to prepare the hides was selected carefully—right down to the type of wood that was gathered, and the location chosen in which to smoke and tan the hide. Even the water that was used to wash the hide was selected accordingly. "It was so amazing how we were connected with nature, with the environment, with the animals. Wherever the animals journey in the winter and the spring throughout the seasons, we would follow and make a cabin and stay there the whole duration of the winter, and in the springtime we would move back down to the river."

It is important for Mary Jane to share these stories, because she says that the knowledge needs to be passed onto the young people. A lot of young people raised in the community are no longer going out on the land. If they knew how to be on the land they would be able to help themselves and their families. Land knowledge is not taught in the school system, but it should be she tells me.

A lot of people who had been living along the river moved

into the community of Łíídlı̨ı̨ Kų́ę́ over time, but Mary Jane's grandparents lived at Fish Lake in the mountain range until the day they died. Mary Jane's late mother told her that her ancestors used to paddle all the way up the river between two mountains, in spruce and birchbark canoes or moose-skin boats. Once they arrived at their destination, they would dismantle the canoes and the women would start making garments with that very same moose hide for the winter. The wooden poles would be put in the trees along the shoreline, so that other families passing by could easily rebuild their boat if needed. "Everything was used. They never wasted anything at all." Zero footprint was left on the earth. Nothing was left behind in the camp when they left; it was either buried or put in the trees. You wouldn't know anyone had been there. This is the way it should always be.

Life on the land back then was not easy, but it was filled with purpose. Keeping active keeps people younger and stronger. Maurice Zoe, like many knowledge carriers, is content with what little he has and speaks of a life well lived when out on the land. While meeting with Maurice for coffee I slowly close my computer hoping he's not referring to me when he says, "Nowadays nobody wants to do physical work. They all want the cyber-world to make money. Press your button here and there. No physical work. They just use their brain to make money. Nobody wants to work hard. They want to make money before they're forty years old so they can enjoy their millions."

"The earth is changing because we are off balance with the spirits."

Maurice tells me that one time he saw David Suzuki at a lecture hall talking about climate change. Maurice says that Suzuki and other environmentalists stand on podiums and make money by speaking about climate change, but they haven't lived on the land and therefore they do not know the effects of climate change as well as those who do. We lived it. We drank water straight from the lake. We slept on the ground. We slept on Mother Earth's breast, stomach, leg, hip, in the bush, and we know what we're talking about, says Maurice. "The earth is changing because we are off balance with the spirits." The land has been cultivated to the point that one day it will not produce any more. In Tłı̨chǫ Yatıì it is called "weka"—we are creating a scab on the earth by scraping the skin.

Now that I am a grandmother, I have a responsibility to try to make sure that we don't keep scraping open that scab. We all want there to be a future to look forward to for our grandchildren and great-grandchildren.

I once asked my best friend Jennie what her dream job was

and she replied, "I can't wait to be an Elder." This is how highly our Elders are revered. I hope that one day I too will be an Elder growing old alongside Mother Earth. I want to embrace that role and the responsibility that comes with it when the time comes, but I fear that I will not know Mother Earth as I do now. That she might be unrecognizable.

— 6 —

Sacred Waters

Water is our most precious resource. Without water we cannot live. It is what gives life to Mother Earth. We are created from water which is why our bodies are made up of mostly water.

Working on the Mackenzie River Basin Transboundary Water Agreement, a groundbreaking agreement that protects NWT waters from contamination, was the first job I landed after graduating from university. It is where I met my best friend Jennie McPherson. We were on the same file together. Getting the provincial government of Alberta to sign the agreement in 2015 was no easy feat, because the agreement prohibited downstream effects from the tar and at the time of signing, the oil industry was at its peak in Alberta, so to have the Alberta premier sign off on the agreement was a victory.

At one of the consultation meetings that we held in community during the making of the agreement, an Elder was given the microphone and said that one day water will cost more than gas—and that day has come. A liter of bottled water can cost upward of five dollars depending on where you live, while a liter of gas is a dollar fifty at the pumps give or take.

One day water will cost more than gas.

Nowadays a bottle of water isn't just water. It often contains other ingredients. Water was never meant to be sitting stagnant in a plastic bottle, or even a treatment plant. It is meant to constantly flow over minerals and pick up nutrients as it travels, so that when it is consumed it fills our bodies with the natural minerals that we need.

In the north we have seen extreme water level fluctuations over the years. Some years there is flooding. Other years there is drought. Chiefs from different nations across the NWT get together from time to time for political reasons, and sometimes they are old friends. Chief Fred Sangris says his friend, the Chief of Tulita at the time, phoned him one spring and said "Fred, I'm ready to walk across Mackenzie River to the other side." He wasn't joking. The river was the lowest it had ever been, sitting at about two feet and getting lower. The barges had stopped running. Fuel and supplies had to be flown into the community.

How could this be? This is the same river that almost swallowed me on a bus ride from Edmonton to Yellowknife. It was at a time when vehicles had to cross over the ice road, before a bridge was built. It was springtime and the ice was rapidly melting on the river, but the bus driver took a chance anyway and drove across the slush. The water was above the wheels and I thought for sure we were going under. We made it across but the bus broke down soon after we got back on solid ground.

Mary Jane Cazon's father told her that wherever he went out on the land he would get fresh drinking water from mossy areas. He would remove the moss from the ground, and immediately after he could see fresh water rise to the top. He told her that one day that would no longer happen because the land would be dry. He blamed it on what was happening down south with resource extraction, says Mary Jane. He told her the water system that runs underground would be affected and we would no longer be able to live the way we used to live.

I learned about the importance of groundwater once while visiting Kátł'odeeche First Nation, one of the only reserves in the NWT. I was sipping tea with an Elderly couple in their living room when they explained how groundwater works. Not much is known scientifically about groundwater, but these two had it all figured out. We all know that water and oil don't mix. Fill half a cup with oil and the rest with water, and you will see that the oil rests on top of the water. Similarly, in nature if you take too much oil out of the ground, or if the permafrost layer above the groundwater melts, the water beneath rises up through the ground until it evaporates contributing to drought and these past few years have been the driest on record in certain places across the NWT, fueling out-of-control forest fires.

The Kátł'odeeche Dene who live along the Hay River have experience major flood events but now parts of the river are just a trickle. It used to be that you would have to drive all the way around the river from one end of town to the other to get to the reserve, but now that there is no water it just takes a few minutes to drive across the bank.

I visit Hay River when I can because my sister lives there. In fact, I spent a month in the height of winter living in a small cabin just outside of Hay River on the shores of Tucho while teaching a class about northern journalism. I forgot to plug my mom's truck in one morning at -50 and it froze like an ice cube. I had to get help to boost the battery.

Even inside the cabin the cold still managed to get in. The wind would howl all night and in the morning a triangle of powder, a miniature mountain, would be piled up in the entryway from the small crack in the door frame.

During that cold spell we had organized a field trip for the class to go out and check nets with the local Elders. My toes froze as we stood under a sun dog on a crisp winter day while one of the Elders showed us how to cut a hole through the ice with an auger and run the netting through to another hole about twenty yards away. When it was time to bring the nets up the next day we had only caught one fish. The Elder said that the fish aren't in abundance like they used to be and they

A giant's face encased in rock, downtown Yellowknife.

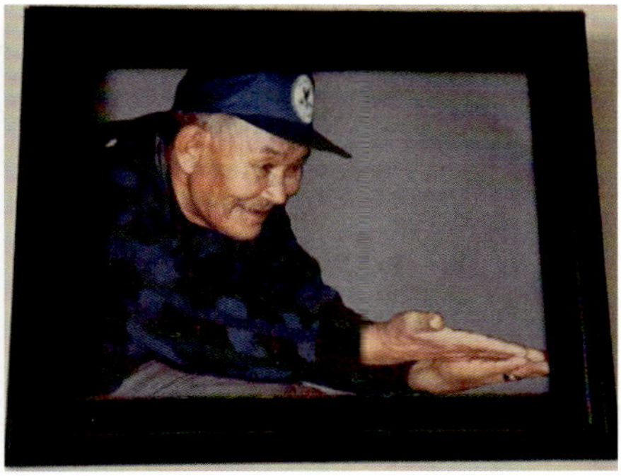

A photo of Alec Rabesca's father, J.B. Rabesca, playing hand games.

An angry fox.

Checking nets.

Dog team.

East Arm of Tucho at Betsı̨ı̨ghıé / Utsingi Point where the water is alive.

Spruce gum medicine, wolverine tears.

Tuktuuyaqtuuq.

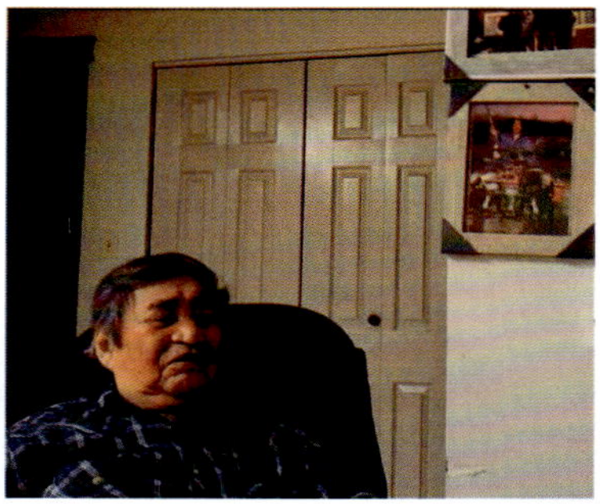

Elder Jonas Noel.

Great-Uncle Alfred Baillargeon.

Iron and arsenic tailings pond at Giant Mine.

My daughter running down the ramparts on the Dehcho at Łíídlįį Kų́ę́ (Fort Simpson), the place where the rivers come together.

Machines that keep down the arsenic dust on the Giant Mine site.

The three giant beaver pelts on Kweteniæaá (Bear Rock).

My canvas tent.

Łutsël K'é puppies.

Helping James cut up the caribou he harvested at the cabin on Sahtú (Great Bear Lake).

Kweteniæ⊆á (Bear Rock).

Jennie caught the sunset in her fishing net.

Lawrence Casaway and his white muskrat.

Lee Mandeville.

On a bush plane to Łutsël K'é.

My son fighting fires.

My daughter fishing in Tucho.

The old church where ʔehtsə́o Erə́ya Prophet Ayah spent lots of time.

Iñuuvik Women's Sewing Circle.

Ehdaalà (White Beach Point).

have changed in appearance due to the effects of the oil sands directly upstream. A direct breach of the transboundary agreement.

With Kátł'odeeche encountering so many significant climate events in the span of a few years, it was almost too much for the small reserve to take. They had been through flood and fire evacuations, a pandemic and severe drought, and as result many people had lost their homes—and when you don't have a home it can become so difficult to cope that it can exacerbate social issues. Hay River has a high crime rate for drug related offences and even murder for such a small town. It's not just Kátł'odeeche, in fact the NWT was listed as the crime capital of Canada more than once and it could be attributed to the hardship of a changing climate.

The next day, as I waited in the airport for the plane, I sat and visited with some acquaintances. We laughed about the big clunky computer that I had to lug around with me everywhere, because I was still working remotely with a law firm that expected me to bring it with me wherever I went. The computer was the size of a small television and slow to start like an old car with an overheating engine and the screen blanked out every once in a while for no reason.

The plane back to Yellowknife was not full. It was only a few people plus a police officer escorting an inmate in shackles. Before takeoff the flight attendant told me I needed to put my computer on the floor in front of me, and I complied after trying to reason with her that it needed its own seat. As we took off, the computer slid down to the back of the plane and landed at her feet. The flight attendant returned it to me thinking it was broken and I hoped it was.

We had to stop in Fort Simpson to pick up a few passangers on our way to Yellowknife. When we landed I got off the plane and put a coin on the ground to pay my respects, because I didn't have tobacco and I hadn't been to Łíídlı̨ı̨˛ Kų́ę́ for a long time. The last time I was there I had driven there from Yellowknife to attend the Dene Nation National Assembly. My daughter was with me and she was quite young at the time, but she was brave enough to get up and sing a song in front of everyone at the local talent show. She didn't win first place but she did win a very stoic looking doll with a headdress. The doll was about four feet tall. She was a bit scared of it and I'm not going to lie, so was I. On the drive home the doll had to stay in the trunk and I don't know what ever happened to the doll after that.

Łíídlı̨ı̨˛ Kų́ę́, or Fort Simpson, is on the banks of the Dehcho. People say that the water in the Dehcho and the Hay River is low for many reasons not just one. Global warming, the dams upstream, the oil sands, not enough snowfall. But now all of these variables are compounding on top of one another, and as a result, of these cumulative effects, the rivers are running dry. If this keeps up, there will be no need for bridges.

And it's not only the Dehcho, all rivers in the north are becoming compromised. As Chief Fred says "The worst is yet to come if humans don't do anything. If they want to save the earth, they have to work fast otherwise the old prophet is speaking the truth." As Fred reiterates this warning, a raven squawks in the background echoing his words.

All the major rivers once had roaring rapids right up until

August, Fred recalls. There used to be a lot of rain in the north. When he was a child all the children in the community used to go out in the rain and dance. That type of heavy rain hasn't returned for almost a decade. If it does rain, it's usually a quick downpour and it's over in a blink of an eye.

The Dehcho and the Liard River, meet just outside of Gilbert and Mary Jane's front door. Directly across from their home is an island called Gros Cap. People would meet and gather there, says Gilbert. There would be a drum dance day and night. From there, some people would go further down the river all the way to the Arctic Ocean and trade with the whalers.

With global warming comes extreme and unpredictable climate events. In a drought the last thing you would think would happen is a flood but that's what happened to the community of Łíídlı̨ı̨ Kúę́ during the Covid lockdown and the entire community had to evacuate. Mary Jane and Gilbert's house is on high ground, so they didn't lose their home like some other people in the community. As a result of the flooding there was a food shortage and the shelves were empty because the barges couldn't run. It took a long time for the community to clean up the mess after the flood and everyone pulled together to help one another even with the difficulty of having to social distance.

These present-day climate catastrophes are a long way away from the days that Mary Jane remembers spending with her father as a child floating in the middle of the river and checking fish nets without a care in the world.

She remembers seeing a layer of film on the surface of the water for the first time as a child. Just past Black Water, there was a creek where she used to pick rosehips with the Elderly women while the men were checking their fish nets. She saw something brown seeping out from the creek. Her father and brother went for a walk to see where the sheen was coming from and they saw a cluster of bent willows. Her brother chopped at the willows until he hit metal. "There was a whole bunch of army jeeps that were buried in that creek." The jeeps were buried next to forty-five-gallon barrels, and the contaminants from the machines were seeping into the soil and water, polluting everything in its wake. The vehicles were left behind from World War II, when the army had been in the area transporting fuel and supplies to their bases. "A lot of machines that they used were never brought back, so they just buried them up here," says Mary Jane.

To prevent our northern waters from being dumping grounds for industry and military, the NWT needs to raise the alarm on the transboundary water agreement to prevent further cumulative impacts in our waters in the north. Even if enforcement mechanisms are put to use, the concern is that punitive law is not working as it should be. Industry can easily pay their way out of fines and not change their ways. Now that the military is starting to turn their attention back to the north for opening the arctic corridor, industry is racing to the rich oil deposits on the ocean floor which means we need to hurry up and bring back our oral Indigenous stories to protect our lands and waters. Indigenous stories can do more to save our planet than any agreement on paper ever can, because Indigenous stories are able to transform hearts and minds. Indigenous stories show us that nature has the right to exist unharmed. We must not forget that just like humans have human rights so does water. Water has the right to be clean and flow freely. During my time working on the negotiations team, we often referred to the agreement as the "kraken"—a giant water creature—because it was such a large, complicated project to undertake. But the transboundary agreement is not the only giant water entity that exists in the north, there's another one much more mysterious and elusive and this one is not found in the boardroom, it lurks in the deepest depths of the Tu Nedhé, the Great Slave Lake.

George Mandeville encountered the large lake creature three times in his life.

I got the percolator figured out and poured him a strong cup of coffee, just as he finished telling me about his near-death experience with quicksand and was about to start in on his lake creature sighting story. "Delicious" he says after taking a sip.

I've heard George tell his lake creature story before, but I ask to hear it again and he retells it the exact same way every time.

The first time he saw it, was in the East Arm of Tu Nedhé, the big lake as George calls it, when he was with his wife and one of his grandchildren on their way home from the Dene National Assembly that was being hosted in Łutsël K'é. It was stormy. They probably should have stayed another day to wait for calmer weather, but they took off back to Yellowknife as soon as the gathering was over following and keeping close to the shoreline. They were halfway to Red Cliff, and two-thirds of the way to ʔetthën Island—meaning "caribou point" in Dënë Sųłıné Yatı̨—in a channel that almost looks like a river. "We were cruising along there, full speed." And that's when he saw it. "To me it looked like an elephant's back. It just appeared about fifty feet in front of me." Either that or it looked like an upside-down canoe he says.

It was about three feet out of the water and about thirty feet long. "I didn't want to run into it, so I cranked the wheel of the boat as hard as I could to the left." His wife started screaming. His grandson was in the back not knowing what the commotion was about. It happened so fast, remembers George. They almost crashed into the wall of the cliff but he had to try to get around whatever it was in front of them. "It looked huge, and I had to act fast otherwise I would have slammed right into it." It was dark-colored. He circled around but didn't want to go back to see what it was because it unnerved him. He quickly explained to his wife what he saw. She couldn't see it because she didn't have a windshield wiper on the passenger side.

I wonder out loud if this is the same lake creature that Tanya Lantz told me she saw, and George is surprised that I know about her sighting. He knows about it too and says that Tanya saw it about three kilometers past the same point just a few days before. Tanya happened to see it on her way to the Dene assembly. It must have been in that area for a few days, George guesses.

"What do you think it could be?" I ask. That's when he starts to describe the second encounter when back out in the North Arm of the big lake. "George Tuccaro and I were fishing close to Utsingi." It was a beautiful day, and "the lake was like glass." George was puttering along at about 10 kilometers an hour, just a bit above trolling speed and Tuccaro happened to be having a nap. George always looks out at his surroundings when on the water, and that's when he noticed something was swimming behind the boat to the left of its wake. George said it looked like it had a large frog head because it had eyes on the top of its head. The creature was so large that its eyes were spaced about two feet apart, and the eyes themselves were about two feet in diameter, like the headlights on a semi-truck says George.

George woke up Tuccaro and asked, "What the hell is that following us?" Tuccaro turned around and saw the water creature keeping up with the speed of the boat and asked George the same question right back. "Pass me my binoculars," George said to Tuccaro. Just as he was passing the binoculars to George, the creature disappeared back down into the water.

In her thirty years of working up and down the Mackenzie Valley, Tanya Lantz has heard stories from the Elders about when the north was a tropical territory. "We had different trees here," she says. The Elders told her it was very warm and that it might be tropical again one day in the north. Some of the Elders she spoke with, who have since passed on, told her that evidence of old trees would be found when they mined the earth and dug into the ground, and they were right once again.

Celine Marlowe has a large photo of her late husband George hanging on her bright green living room wall. George once told her that a long time ago he used to hear the Elders say that the north was once a jungle and that one day it will probably be like that again. Celine hopes it won't be in her lifetime. She would take the cold over the heat any day. Celine tells me that her husband was working at the Ekati diamond mine—Ekati meaning "fat lake" in Tłı̨chǫ Yatıì—when they found an ancient tree. It was way, way underground under many layers of earth.

Rumor has it that when the driver of the excavator drew the claw back and exposed the earth, he had a strong whiff of cedar and got out of his machine to take a look. That's when he spotted an ancient tree stump. The tree is confirmed to be a 50-million-year-old fossilized piece of redwood cedar located 315 meters below the earth's surface to be exact, preserved to near perfection. It was tropical here in the north thousands of years ago and our ancestors knew this. How could they have known this without someone having witnessed it and passing on down the line? Science says that humans are only roughly 300,000 years old, but the findings of this tree and the knowledge that it was there as told by Dene Elders is persuasive evidence that we are much older than we have been made to believe.

With all that we know about the history of the north once being a tropical oasis, it's quite possible that the lake creature could have survived that time period. Whether it be tropical or prehistoric, George Mandeville is not entirely sure that what he saw that day in the water was the same creature he saw the first time, out in the storm with his wife and grandson, but he's not ruling it out.

George tells me that he talked to a man who had employed two divers to search for a body in that same area, after a person fell through the ice and drowned. When the divers came up, they were adamant that they weren't going to dive back down. "They saw something huge that bumped one of them." The divers described it as the size of a whale. "They figured it would circle around them and take a bite out of them," George says with a laugh.

It makes me wonder how many sightings there have been over the years. Even one of the community priests saw it. The late Father Jim Lynn said he spotted it just off T'èʔehdaà Point, not far from the little church where he conducted mass every Sunday. "It's been reported for thousands of years passed on. Every generation has seen it," George says. He has spent a lot of time on the big lake, which makes it more probable for him to have seen it more than once.

There are canals and underground tunnels underneath the lakebed. There is also a theory that the two great lakes—Great Slave Lake otherwise known as Tu Nedhé, Tucho or Tindee, depending on what language you prefer to use, and Great Bear Lake, Sahtú, are connected through underground tunnels. George says that the two lakes were once joined. "It was the biggest lake on earth."

I believe that the two lakes in Dënéndeh are still connected through tributaries and groundwater. Could it be that the giant living lake creature is an inhabitant of both the NWT's great lakes?

The third time George encountered a large unidentifiable creature in the water, his wife and her friend saw it. They thought he'd run over a reef that appeared out of nowhere but it was there and then gone again. The depth finder in that area read 600 feet—they were just before the turn at Devils Channel. It is in that area where an offering is often made by the Dene to cross the lake safely, and the creature that George describes has been spotted in that area several times by many people according to George's friend, the late Alfred Lockhart.

Tucho has 300-foot cliffs sheathed in the world's oldest rock, and known depths of up to 2,000 feet. Take away the water and you're standing on top of a mountain. But there are also parts of the lake that are extremely shallow. George tells me that he measured the North Arm of the lake once when he was guiding a team of biologists who were collecting soil samples. He tied a string to an axe handle and lowered it off the side of the boat. It was eleven feet deep in all directions consistently across the whole bay. Even though it is only a few feet deep, the current is strong and can sweep a person away if they go overboard. This could explain why no sightings of the lake creature have ever been documented in the North Arm area—it's too shallow for a creature of that magnitude.

Every time I hear an account of the lake creature being spotted it is in deep water, except for the time Tanya Lantz saw it. Before she relives the story of what she witnessed in the water, Tanya wants to make it very clear that "not all stories are meant to be shared." Our Elders used to tell us what stories we could tell and what stories we shouldn't share to just anyone because not everyone will respect them. I am thankful that Tanya was willing to share with me what she saw that day in the water. It is important that we tell our stories on our own terms when the timing is right. For far too long, our stories have been culturally appropriated. But how do we know which stories are to be shared and which ones aren't? Dr. Redvers makes a valid point when she asks: "At what level of crises do we need to be in before we decide to break protocol. Do we break protocol in those instances or do we not?" That is the kind of question that we are grappling with as Indigenous people. So when do we pull the trigger on sharing our most sacred stories in order to help tackle the climate crisis? I am treading very carefully in unchartered waters.

The only ways to know which stories can be shared are by asking permission from the storyteller first of all and secondly by creating more opportunities for dialogue on when to expand the threshold of protocol and sharing, says Dr. Nicole Redvers. Now, in our time of crisis, it is time to open that dialogue a bit more, and so Tanya has so graciously shared with me her story trusting that I will carry it with care.

It was a couple weeks into the summer. There was a Dene National Assembly scheduled in Łutsël K'é. Tanya and her group were traveling by boat from Yellowknife to the gathering. The day before they traveled the lake ice had broken up. Tanya had hired a community guide to bring her, her daughter, her friend, her colleague and her colleague's daughter by boat. There are different ways to travel to Łutsël K'é. She chose to travel through the islands.

"As we were getting close to the community, I decided that it was time for lunch. We were hungry and wanted some lake trout. We drove into a bay. The water was quite shallow. It was a super-hot day and the lake was like glass. We fixed up our fishing rods and cast into the water, and as soon as that hook hit the water we got a trout. Not just a small trout but a good-sized trout. We pulled it in. We were excited and happy we scored some lunch. There were two children in the boat so we cast another line for the other child hoping we would catch another one, and as soon as that hook hit the water we caught another really good-sized trout. We did it again a third time just to see if it would happen again and it did. As soon as that hook hit the water we caught another big trout. At that point I turned to the guide and said what's going on? There shouldn't be big trout in a shallow bay in this hot water. Something's chasing them here and we're in the middle of it. He agreed and said it's not quite normal."

It was then that Tanya suggested they go to the shore and have a shore lunch and be on their way, but after that third trout was caught she had what she describes as a heightened sense of awareness. An awareness that they were not alone, that something was there with them. Tanya chalks it up to a woman's intuition. "We went to the shore and the guide that was with us went to the tree line to look for dead wood to start a fire. My daughter was still on the boat and that was when Tanya heard the sound of rushing water. "I never heard a whale spout before... but it sounded like a spout."

She turned around to look at the lake and saw a huge creature as big as a house. It was dark and not far from shore. She says, "Everyone froze for a minute and they were all staring at it and couldn't figure out what it was. The guide came running down the hill and he was ready to push the boat out. I didn't think it was a very good idea. That if we had the privilege and the honor to see something like this it wasn't for us to go investigate further." Tanya did not want to bother it and convinced the guide not to go out onto the water by boat to get a better look. "It was way bigger than the boat and we had children with us. So we stood on the shore and watched it swim in between two points until it went away." The creature was so big it created its own waves. "When it was gone, we finished our lunch and waited before taking off on our journey. Until we felt good and safe to go."

Before that day, Tanya had only ever heard of one other person who saw the water creature—or "water spirit," as she refers to it—and that person is the late George Marlowe from Łutsël K'é, Celine's husband. He told Tanya that when he was a young boy he saw it and the Elders at the time told him not to bother it or investigate it. He informed Tanya that it usually means something in your life will happen one way or the other.

There is talk of trenches, tunnels and underwater caverns in the lake, says Tanya. "I really do believe now that this creature or spirit lives down there." No one knows how far those passageways go, but Tanya recalls Elders talking about tunnels. She often wonders about what other people's experiences in that area near Betsų̨ghıé. "It's a very special corner that you have to go around. The waves go straight up because they come from both sides. It's very scary. If the water is not calm people go around it.

Tanya knows about the divers that George spoke about as well who, rumor has it, even quit diving indefinitely after the incident. A couple decades ago there was a tragedy there, she tells me. Someone drowned. They sent drivers to retrieve the body. When the divers went down they saw something much bigger than them and they never dove again. "That was the last dive that they made in their lifetime. The body was never retrieved."

Tanya herself remembers being anxious about the water. "I didn't go out on the lake for a little bit after that. I didn't go out on the lake without having the deepest respect every step of the way." Tanya says it took her a few years to process what she saw that day. In the north, news travels fast. "When we got to the community, the story got out of what we witnessed."

Because Tanya had built up a relationship with the Elders, her story carried credibility among the community—but there was a reporter there that scoffed at the story, she says. It didn't bother Tanya as much as it bothered the Elders. "I'm not here to convince anyone of what I saw," she explains. The reporter was not from the community and didn't have the respect that she should have had. The Elders didn't appreciate the reporter's blatant and outspoken disregard and because of that the reporter wasn't welcome back to the community Tanya says.

Steven Nitah who was the Chief of Łutsël K'é at the time of Tanya's sighting says, "There are places where we have to be very respectful and quiet and feed the water. There are places where sightings have taken place for generations." The creature has been seen. Photographs have been taken. The Elders said that any photos of the lake creature should be destroyed because they didn't want the outside world to get excited about it. So why then am I telling this story? I ask myself for the hundredth time and all I keep coming back to is what might happen to that water spirit if the lakes continue to recede or become contaminated. Would the creature die and wash up on shore?

If you ask Alec Rabesca he will probably tell you that the lake creature is a mammal. He and his brother were traveling by skidoo on Stark Lake when they came across a large gaping hole in the ice where thousands of shards of candle ice bobbing in the water in Tinde'e, near Łutsël K'é. The hole was about ten feet in diameter. "I'm pretty sure something was there." Alec recalls that it was as if something came out of the water and broke through the ice and then went back under. "I got an eerie feeling when I went close to it, so I didn't want to hang around." Could it have been the same water spirit that Tanya and George saw? If it has to come up for air to breathe as mammals do, then it would have had to break through the ice. And in that case, it would make sense for it to live in areas of the lake where there is open water year-round—and there are portions of Stark Lake that are open all the time. Stark Lake is connected to Tinde'e through a small river system that the water creature could easily swim through right down the hill from the Łutsël K'é airport.

The only prehistoric amphibian I could find that might be close to what George, Tanya and Alec describe is a Phiomicetus anubis, that's a mouthful to pronounce but the simple meaning of the term is a prehistoric whale. With the detail in Tanya's account about the sound of a spout I wouldn't be surprised if the creature had the features of a whale—which could explain why it would have to come to the surface from time to time to breathe. There is also the frilled shark that has been found alive off the coast of Portugal. It has 300 teeth, can grow up to seven feet long, has a fin at the very end of its tail and moves like an eel. This 80-million-year-old species has somehow survived mass extinction events so it is possible that our lake creature is also prehistoric.

Will world-class fishermen be even more attracted to the north after hearing about the lake creature? Would they be foolish enough to try to bait it? Possibly, but they would be hard-pressed to get past the Thaidene Nëné guardians—also known as the Ni Hat'ni Dene Rangers. Ni Hat'ni Dene means "watchers of the land" in Dëne Sųłıné. The rangers assert their Dene rights and authority in the park through their presence and activity on the land and water, maintaining the integrity of their cultural sites and the environment forever.

Maybe the water creature rising to the surface is trying to tell us something. Is it trying to warn us? Should we be honoring and celebrating the water spirit and not be afraid of it? Or could the creature perhaps be a giant beaver that wasn't killed off by Yamǫǫ̀zha? Could the sound of the spout actually have been the sound of a beaver tail slapping the water? When a beaver slaps its tail it causes a splash, and beavers swim with their eyes above water—just as the creature's eyes were above water when George spotted it. Either way, it would be wise to keep our ceremonies about the water spirit alive because our Diné relatives in the south believe that when the ceremonies stop the world will end.

Who would have known that in working on the transboundary water agreement I wasn't only helping to protect pristine northern waterways, I was also helping to protect an ancient water spirit?

The water agreement should have included the words of the late great Dene prophet Ayah from Délı̨nę, who warned that the Sahtú would be the last freshwater lake in the world and that people would flock to it from all across Mother Earth for refuge. He envisioned many people along the shoreline of the lake.

Could Ayah have also known what the lake looks like from above? I have asked around about why the lake is named Great

Bear and the most obvious response is that there are lots of grizzlies and black bears around the lake. The other answer is that it is shaped like a bear from above. I can see how it might look like the hide of a stretched bearskin on a map, but in the days before maps how would the Dene have known that it was shaped like a bear from above? My friend Jennie says they most likely used medicine power to transform into a bird to see the lake from above.

I could not write about Great Bear Lake without first actually being on it, and that is why I went with Jennie and James on their annual fall trip to the cabin. Being gifted a caribou was just the first of many gifts from Creator that I received on that trip.

Getting out to the cabin was a journey in itself. I had to fly from Yellowknife to Le Gohlini, or Norman Wells, a town that was built solely for the purpose of extracting oil. Jennie and James live in Norman Wells but Jennie is from Délı̨nę and James is from Tulita. Both communities are located in the Sahtú Region just down the river from one another and their cabin is located in between their birthplaces.

When I first arrived in Norman Wells, Jennie gave me the grand tour, which only took a few minutes because the town is so small. Then we went for a nice hike through a path of lined birch trees. The next morning, we were to travel the Mackenzie River, the original highway of the north before roads. We would travel up the Sahtúdə́ meaning Bear River, which leads into Great Bear Lake where their cabin is located.

It was a beautiful day for a boat ride. Before pushing off from the dock, Russo, their dog, that looks more like a bear than a dog, decided to go for a plunge. He jumped in the boat and gave a good shake that we couldn't escape, blessing us before heading down the Dehcho. "Russo." Jennie cried and we all laughed. I looked back at Norman Wells as we left the shore and saw three islands in the middle of the river, Jennie said they are artificial islands built by the oil and gas company to extract oil out of the river.

We were warned that the Dehcho was extremely low, and it was. There are parts of the river less than a few feet deep and it's possible to get stuck up on a sandbar. Because of this, James drove the boat very slowly just in case but he didn't mind because he was also on the lookout for moose. A large rock in the distance started to become larger and larger and when we got right in front of it the magnitude of it was spectacular. It was Kweteniææaá—Bear Rock. The place where Yamǫrıa chased the giant beavers, killed them and dried their pelts in the sun. The three giant Tsàcho pelts are a different color from the rest of the rock. They are red and they face the community of Tulita.

The rock is quite massive, and sometimes you can see musk ox climbing its jagged surface. At the very top, there is a large cave that looks like an eye looking back at you. A bear must have once lived in there or maybe still does. Jennie and James sometimes take tourists for a hike up that rock.

Just before Tulita we turned left onto Sahtúdǝ́. "Bear River has a different feel hey?" said Jennie, and I nodded in agreement. It was clear with a rocky bottom, not murky and silty like the Mackenzie River. When I asked her where we were on the map, she pulled up her sleeve and showed me a map of the lake tattooed on her forearm. There's no need for a paper map or GPS with Jennie as your guide.

We came upon Bennett Field, an open meadow with a backdrop of vibrant yellow trees. It is where Jennie and James got married. Russo jumped out of the boat and onto the wooden board that served as a plank, but he nearly lost his balance and almost fell into the river. James refueled the gas tank and we were on our way again in no time. James and Jennie had warned me that the Sahtúdə́ rapids were just around the corner and to brace myself, but I didn't really know what to expect. All I knew was to try not to be alarmed if we hit a rock or two.

I'd never been river rafting before but I'd been down rapids in a lazy river, so I didn't think it was going to be as intense as it was. Before entering the rapids Jennie and I switched spots, I sat in the front and she sat in the back on the makeshift bed that James made for us to be comfortable. James had to stand up and stick his head out the wind flap to see in front of him so as not to hit a boulder, and I can't be sure but he might have had a cigarette hanging out the side of his mouth. When I looked through the clear plastic window all I saw were huge pink boulders everywhere in the water. We were literally skimming the surface of the water at top speed. Only a speedboat traveling light would make it through those waters. I felt like we were hovering. It seemed impossible. We were only in a few inches of water in some spots. James's depth finder wasn't even picking up a measurement, because when the water is that shallow it doesn't work.

I held onto the bar at my side tightly and wanted to cover my eyes and tuck my head into my life jacket so I couldn't see. I was afraid we would hit a boulder and go flying. I looked over at Jennie who had a very serious look on her face staring straight ahead. I wanted to hold her hand like the time we went to a trendy Nordic spa in the city and cold plunged into a large tub of ice-cold water for a minute straight. If we were to hit a rock and go overboard this would have been a whole different kind of cold plunge experience. Jennie saw the fear on my face and mouthed the words "it's okay" and I tried not to move so as not to rock the boat. I felt like I was in the movie *Jungle Cruise.* We were in a no-man's-land without any cell phone service, completely secluded. To top it off we were suddenly under an eerie dark cloud and the rain set in around us helping to put out the small smoldering fires on both sides of the river that were still burning from that summer's forest fire season.

The rapids went on for a very long time. We were traveling at about forty kilometers an hour for twenty minutes straight without slowing. When we made it through the rapids the sun came out again, and we continued on at a reasonable pace the rest of the way to Sahtú through a winding channel. Jennie and I took a big sigh of relief that we made it through as James carried on like it was an everyday occurrence. The man deserves a medal.

We came to a small sign in the trees at the end of Bear River with an eagle perched just above it. The sign read "Sahtú" and pointed to the lake. The eagle flew low in front of us once it saw our boat—a greeting perhaps, a good sign. On land I shook my head in disbelief that we had made it, and Jennie said, "I'm not sure how much longer we'll be able to do that." With the water setting lower and lower each year it may not be long at all. Rapids are the filter of rivers and lakes. They breathe life into water. Without them, water becomes stagnant. I'm afraid that with the low water the Bear River rapids will be no more which doesn't bode well for the connecting tributaries. Not being able to cross the rapids severs the ability for the Dene in the area to travel through that area and with the water levels getting lower each year there's more chance of an accident as well, which can easily become a life-and-death situation—especially when there's no one around for miles and miles to come to the rescue.

We didn't have to travel on Sahtú for very long once we came out the mouth of the river before we reached Déline̜, with the sunset in our rearview. I was finally on Great Bear Lake. I had been looking forward to this trip for a long time.

When I stepped foot on the dock in Délı̨nę I took out some dzǫdiì, muskrat root, from my medicine bag and placed it in the water, to thank Creator for a safe journey and two little muskrats swam up next to it thinking it was food. At the same time, I overheard James and an Elder talking at the boat launch about how someone recently went across the bear river rapids and hit a few boulders.

We overnighted in Délı̨nę before heading to the cabin. That night I slept alone in a bed and breakfast. I had two vivid grizzly bear dreams, back-to-back. Whenever I dream of an animal it is usually a bear which leads me to believe that I am related to the bear. Délı̨nę is known to be a very spiritual place and every time I go there it feels as though there is a strong spiritual energy. The first dream I had was that I was out camping with my grandma and a big grizzly came sauntering over to us. I stood in front of my grandma to protect her. The bear had a log in its mouth. I grabbed the log out of the bear's mouth and threw it and the bear fetched it like a dog. There was another grizzly to the right of us where the log landed, and the two grizzlies became distracted by one another which gave me enough time to run. I wanted it to chase me so it would leave my grandma alone and the bear did follow me. I pulled out the knife I brought with me for the trip in case the bear attacked me but it never did. Then I woke up and realized that it was my grandma's birthday. She had come to visit me in my dreams. It was the middle of the night so I fell back asleep and that is when I had another bear dream. This time I was in a busy restaurant and I was outside looking in through a big glass window. There was a bear outside dancing—it was trying to dance like Elvis but didn't quite have the moves. I thought it was me for a moment but there was also a cub next to it. The larger bear was shapeshifting, morphing into a human, in front of my eyes. It seemed to be trying to get the attention of the people inside the building but no one cared. It was like it was trying to entertain them.

I told these dreams to Jennie and her aunt in the old church where Prophet Ayah spent time, before heading back out on the lake the next day to the cabin. Her aunt warned that I should not talk about bears while out on the land because they can hear when they are being talked about and will come around and so I never spoke of bears again that whole trip.

On our last day at the cabin before packing up the boat all three of us and Russo had walked to Grannies Point. On our way there I found a flat stone and tried my luck at skipping it across the water. I've never been good at skipping rocks. The most I might get is three skips if I'm lucky. Well, for some reason I got about ten skips and was super impressed with myself. On our way back, in almost the same spot that I found the flat stone I found a loon feather. I picked it up and brought it back to the cabin and flattened it in a book. I knew the feather carried a message but I didn't know any Dene stories about the loon and hoped to learn more. I got the answer I was looking for when visiting with Elder Lawrence Casaway a few days after getting home from Great Bear Lake when I sat with him on his front porch with the sun shining down and the sound of wind chimes blowing in the breeze. I had not told him I found a loon feather and was hoping to find out what it meant. It was like he read my mind because the last story he told me that day was a story about a loon.

Lawrence, like me, was raised by his grandparents. He wanted to tell me the story that his grandfather had told him in Dëne Sųłıné because it is better told that way. Indigenous stories are much more powerful when told in the language because there are words that English just can't describe, but because I cannot understand my language—through no fault of my own, he warmly reminds me—he resorts to telling me it in English.

"Loon has a good heart. Loon was once pitch black, no markings." One day Loon came across an Elderly man who was blind. The man was sitting on a rock by the water and Loon could see that the man was sad. The man was listing to the birds in spring and all the sounds around him. Loon came up to the Elderly man and asked, "Why are you sad?" The man said, "I can hear you but I can't see you." Loon then said, "Hold onto my wings," and the man did what Loon instructed and held onto his wings and Loon dove down into the water. When they came up to the surface Loon asked, "What do you see?" The man said he could only see a blur. Loon dove down under the water again with the man and when they came back up Loon said, "Now what do you see?" and the man said, "I can see a little bit." Loon dove once last time and brought him to the rock he had been sitting on. The man opened his eyes and cried, "I can see!" Loon made a beautiful calling sound and flapped his wings in the water out of joy. The Elderly man was so thankful to Loon for giving him the gift of sight that he wanted to gift Loon something in return, but all he had was a bone necklace. He tossed the necklace out into the water where Loon was and it landed on Loon's neck and circled around it a few times until it settled. That is why Loon has a beautiful design on its neck and is more beautiful than all the other winged ones on the lake.

On the boat ride back to Délı̨nę we cruised under a blue and pink sky in the wake of a boat that was long out of sight. I pointed to the ripples in the water and over the sound of the loud engine Jennie said "water holds memory." I couldn't help but think that if Ayah's water prophecy were to come true, at least I had a friend for the end of the world.

"Water holds memory."

— 7 —

The Land Is My Classroom

In raising my children, I've tried to instill in them a sense of cultural pride in who they are and where they are from. I wanted to give them what I didn't have growing up. As a young man, my son had no previous experience of going out on the land, yet he instinctively knew how to set snares, gut and scale fish, and run a dog team. Knowing how to be on the land is second nature to him. It's in his DNA.

Once, when he and I and a group from Dechinta were stranded on an island, with the wind raging and glacial water bashing our boat into the rocks, we huddled in a circle waiting for the weather to calm. We weren't prepared to be caught in a sudden storm. We were completely exposed to the elements and vulnerable but my son, who was just a young boy at the time, did not cower. Instead, he got to work helping to start a fire to keep us warm and picked wild plants to make tea to keep us hydrated. Later that evening, when we made it back to camp safely, we thanked Creator. We prayed together in the dim-lit wood sauna for the health of our families and loved ones. We prayed for peace, love, happiness and good health as the ancient songs passed down by our ancestors rose louder and louder out of the drummers whom my son admired and looked up to. The beat of the drum kept pace with our own hearts and the steady rhythm resonated deep in my cold and achy bones. The land teaches us to give thanks. Now my son is a father and his sons play the drum. When my first-born grandson drums he often stops and points up to the sky with his drum stick. Children are much closer to Creator and can see things that we cannot.

Being out on the land can teach us all the same subjects in school. You want to learn about history? Listen to our ancient stories. You want to learn about science? Go and sit under the northern lights. You want to learn about biology? Learn how to gut and scale a fish. You want to learn about geography? Walk along the shoreline and collect shale. You want to learn about nutrition? Go out with an Elder to harvest plants and medicines. Learning on the land is much better than sitting in a stuffy classroom. Being on the land is hands-on learning. What students are learning in school these days is not useful in the real world. Students need to learn about things that matter, things that can be applied in their day-to-day lives. Students should be given the space to have meaningful conversations about how they can influence change in the world not sitting around all day learning about how the explorers colonized the Americas. The entire education system needs a rehaul. Students should be learning about their futures. Being on the land teaches resilience, it builds strength and character, it is good for the mind and body and helps alleviate anxiety which is something that a lot of young people are experiencing these days. Being on the land teaches us that everything is intricately connected and that just like the ecosystem we are diverse. This intersectionality can help students tackle life's problems with courage, wisdom and confidence.

My cousin Randy is a father and he is raising his son out on the land as much as he can. "He asks a lot of questions," says Randy with a laugh. I say that's a good thing. We don't ask enough questions as adults; many of us have lost our childlike curiosity.

Ahfrè is learning about the world—his world. Last summer Randy and Ahfrè checked nets almost every day after breakfast. They boat around all the bays and islands close to T'èʔehdaà, and Randy points to and says the names of the places they pass by in their language, so Ahfrè will learn them. They usually stop and have a snack in one of the bays when checking the nets. One day Ahfrè pointed to a big eagle in the sky and said, "Det'ǫcho." He is learning to speak his language. On their way back home passing by the islands and bays, it is Ahfrè's turn to point and say out loud the names of the landmarks that his father has shown him. Ahfrè is only three, but he knows his place in the world.

"I remember I did that with my grandpa. I used to check nets with him. That's how I know the boat trails." When Randy first took his boat out on the water alone, he was scared of hitting a reef, but now he's not. He finds it funny that people use GPS. He took his friends out on the lake once and they seemed worried that he didn't know his way. He took the inland trail when most people take the open route because it's not as difficult to navigate. It's easy to get lost in the islands inland.

His friends pointed out the crosses in the distance that they passed by. Randy had to explain to them that those crosses marked the places where Dene families lived before their children were taken by the church—some of them to never see their families again. Sometimes people have to see it to believe it, he says.

Like all Indigenous nations, the Dene kept their stories underground at a time when oppression and the attempted eradication of their entire identity was at its height—during the days of residential schools, when a severe effort was made to "kill the Indian in the child," a sinister attempt to reduce the Indigenous worldview into a capitalist one. For over 100 years, residential schools were a genocide mandated under the Indian Act in Canada. If Indigenous children did not attend residential school, their parents would be jailed or social assistance would be withheld. Some Dene families who lived out on the land outside of mapped communities remained out of sight of the missionaries while other families fled into the bush to hide so that their children would be spared from having to attend. Many children that were taken to the schools never returned home, having died at the hands of missionaries who severely abused them in every way imaginable. Either that or they died from the flu epidemics that swept through the inhospitable institutions.

The residential schools were designed to force Indigenous people to let go of their beliefs and to brainwash them into thinking that medicine power is evil and that their original language should be abolished. The schools, run by both the church and the Canadian government, were structured in a way that would have Indigenous children conform to a dogmatic belief so far removed from their own that it would make it easy for the government to systematically take control of

Indigenous people's land instead of having to uphold the treaty promise to live in unity. To shed light on just how wicked and racist the residential schools were, and how prominent an idea they became among colonizers wanting to dominate, the residential school system was later adopted as a model for the South African apartheid.

But despite the hardships faced in residential schools by the church and the oppression wielded by other areas of government, we have still miraculously held onto our belief systems and our ancient stories. Some of the stories are fragmented, some of the medicine is forgotten, but it is still there. Patiently waiting to be rediscovered.

Even now the education system does not make much room to learn from different perspectives; that is because the education system was originally meant to brainwash students to conform to a biased narrow-minded worldview. The Elders say that we, the Dene, must be strong like two people and walk in both worlds, meaning that in order to get by in modern times, we must be educated in both the white man's way but not lose sight of our own worldview.

It's people like Steven Nitah who exemplify what it means to walk in both worlds. "I was raised on the land with Elders, I was taught a lot of what I know from them. I participated very little in school," he says. Steven entered kindergarten at the age of ten and only went to school three months out of the year, because most of his young life was spent out on the land. In the past kids were not allowed to be absent from school to go out on the land, but now there are exceptions and the school system in the north allows for cultural days because the Indigenous leaders pushed for change.

"We used to go to the bush in mid-October," Steven tells me. He would go back to the community for Christmas and attend school during the winter months, then head back out on the trapline in the spring. "I'm not conditioned by the education system to just accept science for science's sake. I have seen value in our belief systems, our value systems and our knowledge systems and I bring that to every engagement I'm involved in."

I first took a personal interest in environmental issues when I went down south to obtain my Bachelor of Arts in Justice Studies at Royal Roads University. I wrote my undergrad thesis on toxic torts, which was driven by my need to understand how it came to be that a gold mine in my backyard got away with wreaking havoc on the land without consequence.

Once I finished my bachelor's degree I moved back home and was voted in as a council member for my First Nation, and we collectively demanded compensation for the arsenic trioxide contamination of our lands from the gold mine. Years later, the federal government has finally agreed to negotiate for compensation, but no amount of money can fix the land that has been left in utter disrepair.

I didn't want to stop learning after graduating so I enrolled in the Environment and Management master's program at Royal Roads University, where I studied the cumulative impacts of environmental racism across Canada. I first heard the term environmental racism when I came across the work of Majora Carter, a Black woman from the Bronx who said "Race and class are extremely reliable indicators as to where one might find the good stuff, like parks and trees, and where one might find the bad stuff, like power plants and waste facilities."

Environmental racism occurs globally in Indigenous communities and stems directly from colonial industrialization for the benefit of capital gain. It is found in the leftover remains of old abandoned mines and mills, in ongoing deforestation, unsustainable corporate agriculture, pipeline spills, fracking and so on. These exploitative eyesores all have one thing in common: they are located directly on or near Indigenous lands and hidden away from the rest of society.

Halfway through my master's program, life got in the way. As a single mother it was hard for me to afford to put food on the table and continue my education, so I took a break from my studies to focus on my career and moved back home to the north. It was then that I was hired by the Government of the Northwest Territories to help create the groundbreaking water agreement with Alberta, British Columbia, the Yukon Territory and Saskatchewan. These jurisdictional cross-boundary waterways are all connected and flow downstream into the NWT's Mackenzie River Basin, which contains our two great lakes. As part of the negotiation team, I traveled across Dënéndeh listening to what Indigenous leaders and community members had to say concerning the health of our pristine northern water basin and the impacts of downstream oil from the bitumen capital of the world, Alberta. Initially signing with Alberta marked a significant benchmark. It was the first agreement of its kind, ultimately setting a precedent, and it is now used as a model for other countries to implement, but it still has not done enough to safeguard our waters because it lacks the spiritual component of our Indigenous stories.

I then moved on to become Director of Indigenous Education for the Dechinta Centre for Research and Learning, a land-based initiative delivering accredited post-secondary education and research in the north. Dechinta, in one way or another, means "in the bush" in many languages across Dënéndeh. When traveling to Ottawa to lobby the government for funding, I was in a taxi and the driver asked me what I did for a living. I told him I worked for a land-based university and he asked what *land-based* meant. He said he envisioned a bunch of people on a small island with water all around us. He asked if we were water people. It was then that I realized that not everyone knows what being out "on the land" really means, and I had to explain to him in the most simplistic terms that it was about spending time outdoors.

Ultimately, I made the decision to go back to school but instead of returning to my environmental studies master's program, I decided that I could potentially make a bigger difference through the practice of law. I applied on a whim to law school, not thinking I would actually get in. I had to pay out of pocket with what little I had to travel to the nearest city of Edmonton, Alberta, to write my LSATS but I knew I didn't do very well on them so when I got the call saying I got in I couldn't believe it. There just so happened to be one seat left after a registered applicant dropped out at the last minute and I was the next one on the wait list. My kids and I once again packed up and moved down south. My son did not like living in the city though and moved back home as soon as he was old enough. He is a northerner through and through.

During law school I had to juggle my studies with a job on

the side to afford to live. I worked as a journalist for a media outlet called IndigiNews tasked with the education beat.

I almost didn't make it through law school but I eventually completed the Indigenous Legal Orders and Common Law Juris Doctor at the University of Victoria. I strongly believed that to help minimize the trajectory of human-made change in my crusade for climate justice I must first be well versed in how legislation, policies, regulation and corporate structures operate, to ignite change—but the truth is that anyone can be an environmental activist and it doesn't take a law degree. Being taught by incredible Indigenous professors such as John Borrows and Val Napolean, was the best thing that happened at law school and I am thankful for that but I did not take away much else. Most of what I learned was either over my head or went in one ear and out the other but when Professor John Borrows taught us about the five sources of Indigenous law I was intrigued because it made perfect sense to me unlike the common law. The five sources of Indigenous law are:

- Sacred teachings (creation stories and spiritual principles)
- Naturalistic observations (animals and non-human life interactions)
- Deliberative practices (council and community discussion)
- Positivistic law (proclamations, rules and codes)
- Customary law (repetitive patterns of acceptable social interactions such as marriage ceremonies and adoption practices)

As you can see these laws are much different than business, property and criminal law. The sources of Indigenous law have meaning, purpose and connection and are not centered on individualistic ideologies like the common law.

Throughout my time in law school, I felt as though I was being molded and shaped into a box I would never fit into. But my friend, a Cree law professor who also went through the hardship of law school, gave me good advice that got me through. He said just play the game—you don't need to believe in it to play it, you just need to know the rules. He knew I was angry when I learned that the common law under the British monarchy was used as a tool to gain control of Indigenous lands and people. The term dominion of Canada did not sit well with me. It's no wonder that property law was by far my worst grade.

Through the common law, Indigenous lands across Canada have been encroached on, displacing and forcing out many people from their homes. The common law doctrine of "fee simple" property, where homeownership is a commodified land grab, was and often still is a violent process for Indigenous peoples. Much of the fee simple land was taken up through outright land theft. Where there were once pristine hunting, fishing and gathering sites for Indigenous people, there are now expensive homes with signs saying "Private Property," "No Trespassing" and "Keep Out"—even though they are the ones who trespassed on Indigenous lands in the first place.

The government of Canada still does not view Indigenous law as equal to the state—or within the realm of international law for that matter. It is difficult and expensive for

Indigenous people to fight for our rights in court. Our oral histories are not often accepted as evidence.

To understand the theft of Indigenous land in Canada, you must first understand Aboriginal rights and title. Aboriginal rights are the rights of Indigenous people to practice their cultural lifestyles such as hunting, fishing, trapping and gathering—and generating profit, if need be. Aboriginal title is an inherent right, recognized in common law, that originates from Indigenous people's occupation, use and control of ancestral lands prior to colonization. Aboriginal title is not so easily granted by the courts; because it is a property right.

Before the Royal Proclamation, Spanish colonizers—armed with their laws and religion—came to North America and declared that the land was *terra nullius*, meaning that there were no humans, when in fact there were millions of Indigenous people living and thriving across the continent. The way that the *Magna Carta* got away with imposing their laws was to claim that Indigenous people were not human. Insinuating that we were animals and therefore could be dominated and ruled over as though we could be owned. This allowed the Church to claim that Indigenous people were conquered as abhorrently set out in the 1763 Royal Proclamation, but we were never conquered. We never waved a white flag. The law of *terra nullius* was officially denounced by the Supreme Court of Canada in this generation but it's impact will still linger on for generations to come.

There was a time when the Crown referred to the northern territories as "Rupert's Land." The government of the day had purchased Rupert's Land from the Hudson's Bay Company.

Apparently in those days the north could be bought and sold before it became official Crown land. Rupert's Land made up one-third of all of Canada's land mass. After the transaction, the government then had the audacity to turn around and claim the lands they had purchased were "exclusive," which is the highest form of ownership under the common law. But the government didn't have the right to claim that they had exclusive title in the north, because if they did, they wouldn't have had to buy it in the first place.

Part of the promise of the first historical treaties between the Crown and Indigenous people were to collectively care for the land as equals but Canada did not uphold their part of the treaty, which is a huge understatement. This grave misunderstanding between opposing worldviews has prompted the need for modern-day treaty negotiations so that the Canadian government can work to reconcile the past wrongs that they committed.

Francois Paulette is a revered Dene Elder for many reasons—one being his work on the *Paulette* case, a legal caveat that ensures that the government keep to their treaty promise that our rights as Dene people are upheld. The promise that the treaties would be honored as long as the sun shines, the grass grows and the rivers flow has been broken because the land and people have been exploited. The treaties that were signed across the north more than a century ago were not an agreement to cede and surrender and give the land over to the government but to rather protect Dënéndeh for all times for future generations.

It's no wonder that after law school I wasn't sure I wanted to practice law after all, the chance to create real change felt hopeless but I went ahead and tried anyway. *What have I got to lose?* I thought. When it came time to write the bar exam there were only a handful of Indigenous people out of the 300-plus people who were taking the bar. I felt alone. I did not do well on my first attempt. I didn't do well on my second attempt either. To be fair, I didn't really study. But it's a "three strikes you're out" game so I had only one more chance to write the exam to pass, the only problem was, I wasn't sure if I wanted to play the game anymore.

The day after I caught the large trout at Jennie and James' cabin, Jennie showed me how to fillet the fish like a professional. She's an excellent teacher because she shares what she knows modestly. There are some things she did not know. For instance, she didn't know how to cut off the head of the fish I caught, and I was no help because I didn't either. This lack of knowledge is not our fault; it is a fault of the interruption of colonization that has causes us to not be able to practice our culture. Indigenous people shouldn't hold shame for not knowing how to practice our ways of knowing and being. Jennie approaches teaching with humility. She doesn't feel like she should be expected to know everything, and it's important to her that other people know that it's okay not to know how to do something that might be expected of us too.

Together we eventually figured out what to do with the fish head and put it with the rest of the leftover parts of the fish and caribou that we weren't going to consume. James put it in a container and tied it to the front of the boat to later feed to the water. Before feeding the water we stopped at an open area where we knew there would be an abundance of berries. We walked over spongy moss, and I pulled some back to check underneath to see if the water would pool like Mary Jane Cazon's father had shown her, but it didn't; it was bone dry. Moss is supposed to be like a giant sponge. The moss looked like hardened coral on a dried-up ocean floor.

The land was so dry that the berry bushes looked like they had been burned in the sun and the berries were few and far between. Mind you, it was fall, but James had said that the season was running late that year. Mary Jane Cazon informed me that the berries were changing, but now I was seeing it with my own eyes. Drought conditions for food sources such as berries in the summer will affect not only humans, but also the animals that rely on food sources that are only available at certain times of the year.

James looked very comfortable lying on a bed of moss with his head propped up on a tree stump and his gun next to him keeping watch for bears while Jennie and I set out to pick berries. I trampled around impatiently looking for berries, taking a little bit from each bush until Jennie, who was already quietly sitting down picking berries from one bush, said that when I find a berry bush, I should thank it and take my time gathering all the berries before moving onto another bush. And so that is what I did. I found a berry bush and sat down beside it. That day I learned to slow down and bow down to the berries and when I did more seemed to appear as I sat in silence and focused.

Earlier that day Jennie and I talked about the water heart being the heart of the world, and as we were picking berries we got onto the subject about what the brain of the world might be—could it be that the brain is our collective consciousness? We of course did not have the answer, but when walking over mushrooms Jennie said she wanted to be more knowledgeable about the different types of mushrooms in the north, and that was when we had the epiphany that maybe mushrooms are the brain of the earth. The mycelium of fungi is like an advanced connective tissue that enables them to communicate with other fungi. I don't know much about mushrooms either. I don't know which ones are poisonous and which ones are safe but I do know that they have a very high frequency similar to flowers. If mushrooms take over the world one day, I'll be the first to say I told you so.

When our bags were about halfway filled with berries, spruce gum and Labrador leaves, we called it a day and got back in the boat and cruised to the opening of the bay, where the waves were so big it looked like the lake was breathing. James steered the boat in line with the waves, and it felt like we were riding the surf. He stopped just before getting out onto the big lake where there was no shelter from the wind, and asked Jennie to take the wheel so that he could climb out the small door to the front of the boat and offer the water the remains of the fish and caribou that we did not use.

It is very important to take care of the bones of animals. If you do this, more animals will come to you for you to harvest, says Mary Jane. She was taught to place the bones in a tree that has large limbs, so that birds and other animals can scavenge more easily. Across Dënéndeh there are different practices when it comes to discarding animal remains. Sometimes we feed the fire, sometimes we feed the water, helping Mother

Earth do what she does best, reclaim—just like she will reclaim our bones too one day.

I had an epiphany after that trip to the cabin on Great Bear Lake, and that is that I am enough. I had gone out in search of answers about what direction I should take in my life—mostly about whether or not I wanted to continue to pursue law—and the answer I got was that I was enough.

For me Creator is everywhere especially when out on the land, so it was not hard to seek out those quiet moments of clarity while out at the cabin. Before the answer came to me, I had not felt so lost in a long time, and I was not sure which way to go anymore, but when I took and the time to fully immerse myself in nature I was given the answer I was looking for because I removed myself from the distractions of day-to-day life. The answer had always been with me. Deep down I already knew which path I was supposed to go down.

As much as I didn't want to leave the cabin, I couldn't wait to take a shower back at the bed and breakfast in Délı̨nę. I hadn't showered in a few days and had been looking forward to having hot water again, but once I cleaned the dirt out from under my nails and washed the smell of woodsmoke from my hair, I felt a sense of longing. I had not worn makeup the entire trip and it was the longest I had gone without brushing my hair. The land stripped me down to my true self. I don't need to be someone I'm not when I'm out in the bush. I can show up as I am, without expensive clothes, a nice car or a fancy degree. I can be my authentic self when on the land without judgement and the need to impress. The land reminded me that life is not about materialistic things. The land reminded me to keep grounded, literally. The land consoled me and gave me permission to love myself, to take off the mask.

If there's one thing I've learned over the years, it's that if I ever have an important decision to make, I seek out nature and ask Creator to guide me to the answer. In trying to become a lawyer I felt like I was swimming against a strong current, and once I decided to stop going against the forces of nature I was able to let go and float gently down the river. When I told my family that I was no longer going to pursue law, most of them said that if I just kept trying I could do it, they encouraged me to not give up. But my daughter said something to me one day that really hit home. It had been a particularly stressful day at work while I was articling, and she could see that I was upset, she looked at me and said, "Why are you even trying to be a lawyer when it makes you this stressed out?" I was shocked and protested, "I'm doing this for you." It was true, I was trying to prove my worth to everyone but myself. I was putting my own dreams aside to become someone I wasn't, so that my family could be proud of me and have someone in their life to look up to. But the truth was that they were already proud of me. They had never expected me to become someone I wasn't. Once I realized that my family had no expectations of me it was like a weight lifted off my shoulders.

For so long I had been standing in the doorway with one foot in and one foot out, and now I was able to close the door on law and follow my true passion without feeling guilty.

It was the right thing to do. I had developed a complete disdain for the law over the years, to the point that I couldn't even pick up a legal document without feeling like I wanted to crumple it up in my hands. I had managed to get through law school without reading a full case. When I tried to study for the bar exam, the text seemed so trivial and full of unnecessary details that I couldn't stop myself from being so bored I'd fall asleep. But writing is where my true passion lies. I'm a storyteller. I always have been. Every waking hour that I should have been studying, I was talking to Elders and knowledge carriers hanging onto their every word in the making of this book. It was where I found excitement. It was what got me up in the morning. Story is where I find passion, purpose and meaning, not in some dusty old book about wills and estates. All I needed to do was to learn to listen to what my heart wanted and follow it.

I have not thrown away that hard-earned piece of paper that says law degree. Those years of blood, sweat and tears were not for nothing. I will still use what I learned in law school to make change. But it doesn't have to be in a courtroom. You'll find me practicing law in story out on the land as often as I can.

— 8 —

Corporate Criminals

Humans have had a tremendous impact on the environment propelled by the industrial revolution. The world would look a lot different today if we had skipped that era altogether and jumped straight into a technological revolution. Today new technologies have opened the door to the sustainable use of natural resources such as wind, solar and biomass, a renewable organic material that comes from plants and animals. These renewable sources of energy align with Dene worldviews, because they do minimal harm to the environment. Advanced technologies that harness nature in its purest form can complement age-old Dene principles. We can now live in spaces that have net zero carbon emissions.

Flushing toilets, the flip of a switch for light and instant hot water are luxuries that really haven't been around for very long in the grand scheme of things, yet we take full advantage of these conveniences and most of us are more wasteful than we realize. When my grandmother was young all she knew was how to live off the grid. Every day she and her siblings would chop and haul wood, chip frozen lake ice to draw water for bathing and drinking. She didn't have a toilet, instead she used a bucket to go to the washroom outside. Her family used natural heat from a stone hearth to keep warm. I would not have it in me to live the way my grandma once lived—she washed her clothes in the lake, cooked over an open fire, slept on a mattress made of spruce boughs and used moss for diapers. She didn't have running water or electricity. She didn't even have a fridge—she stored her food in the cold, hard ground. She would often tell me about what it was like living on the land, it wasn't easy but it was rewarding just the same. I know I cannot live like my grandma once did, but I might just be able to meet her halfway using today's technology.

Living off-grid as an act of resistance. As a reclamation of sovereignty, as part of the Land Back movement. Land Back is a grass roots Indigenous-led movement that has been around for a long time gaining movement through campaigns for land reclamation and speaking out against historical dispossession globally. Many of the communities across the NWT live in housing but the housing system in all of northern Canada is in shambles because the units were built by the government with cheap materials. Many people are living in dilapidated homes with mold in the walls and poor ventilation. The government is only now talking about fixing up these houses but when they do they must not make the same mistakes. Houses in the north need to be built to withstand a tumultuous climate in the years to come. The government cannot keep building housing units without factoring in climate change especially with growing concerns around permafrost melt and the onset of coastal erosion, New and innovative home designs retrofitted for off-grid cultural lifestyles is sorely needed.

In the north it is very expensive to live. High costs of rent paired with high utility bills makes it nearly impossible to save money. Right now, the average cost to heat a three-bedroom home in the winter in the north can cost up to $800 per month. As a single mother my power was once cut off for being only a few days late on my bill payment. Thankfully it was in the summer, but I know people who have had their power turned off in the dead of winter for fifteen minutes at a time. It is illegal for power corporations to turn off someone's heat in the north in the winter for not paying their bill, so the power company found a way around it by flipping the switch in fifteen minute increments—just enough to cause discomfort but not long enough so that the company doesn't have a human rights complaint on their hands.

When Dene lands were encroached on we didn't have much of a choice but to become dependent on the use of expensive fossil fuels to keep us warm because woodstoves were replaced with a tank full of expensive diesel, which is now the main source of heat in all NWT communities. The NWT has been blamed as being a serial emitter of diesel, but Daniel T'seleie wants to set the record straight and provide context as to why that is.

Daniel is a retired Dene lawyer, but he is still young. Like me, he decided being a lawyer was not for him. Daniel now does consultation work for his hometown of Rádéyı̨lı̨ Kǫ́ę́, otherwise known as Fort Good Hope. He works in the area of Indigenous rights and environmental issues. Daniel was born in Yellowknife but Rádéyı̨lı̨ Kǫ́ę́ is his home community because of his lineage.

"It's true that, a community like Fort Good Hope, which has diesel generators for home heating needs, has high per capita emissions, but in terms of absolute emissions we are a drop in the bucket," he says. Daniel, who is on the board of Indigenous Climate Action, an Indigenous-led organization that is guided by a diverse group of knowledge carriers, water protectors and land defenders from communities across the country, brings his knowledge of the north into all that he does, and generally tries to make informed decisions that reflect where he's from.

About half of the emissions produced in the NWT are from extractive industries not from the communities themselves. The NWT is emitting less than one megaton of carbon a year, which is somewhere around one percent of the emissions produced by the extraction of bitumen in the Alberta tar sands. He stresses that the extraction of synthetic oil in the continental pipeline network is potentially emitting 100 times more emissions annually than all the people in the NWT combined.

Ultimately, it's the large agricultural and industrial emitters, the army, government operations and fossil fuel companies that have the potential to make or break us in terms of reaching the tipping point of catastrophic climate change, says Daniel.

Carbon emissions stay in the atmosphere for thousands of years. All the emissions released since the start of the industrial revolution are still up there, and they are going to be up there for at least another half-millennia. The bulk of those emissions come from globally developed nations that are predominantly white he explains. That is the demographic that is most responsible for the carbon currently in the atmosphere and the climate change we are experiencing. That is also the same group that has disproportionally benefited from those activities. "They took up all the global carbon budget and got rich in the process, and now they want to erase that history and act like it's a clean slate and we all need to reduce emissions equally across the board—and that's just patently unjust."

"We have undergone a century of colonialism that has brought us to this time."

Not only have they gotten their fair share of the carbon budget, they are now better able to implement an energy transition with solutions to climate change because they have the money, he says. They are getting away with not having to face the obligations they have to fix the problem they perpetuated. The blame and onus should not be on communities like Rádéyı̨lı̨ Kóę to somehow figure out how we are going to move away from diesel emissions, says Daniel. "We have undergone a century of colonialism that has brought us to this time." In terms of the wage economy, "we are disproportionally poor compared to other populations in Canada. We don't just need fossil fuels to live and stay alive in the winter" says Daniel, "we rely on them to maintain that connection we have to the land."

Our language is land-based, "people need to go on the land to speak the language and learn the culture and the only way to do that is to use modernized methods of transportation since our traditional ways have been replaced," explains Daniel. "There's a story about a bone in the neck of a duck or a goose that looks like a man. You have to be out there hunting and cutting up that bird to see that bone to tell that story. You actually need to be out there on the land to practice and pass on our language, culture and spirituality. Those are basic international human rights, and right now the reality is we need gas to do that, with skidoos and outboard motors."

There needs to be a justice-based analysis when looking at who should be accountable, says Daniel. The responsibility to swiftly transition away from carbon is "on the fossil fuel companies who have made billions of dollars of profits a year, and have been very actively lobbying against climate science and against real solutions for climate change for decades. They are the ones who should be bearing the cost and hardship and the heavy lifting that needs to happen immediately."

Anything we do in a place like Fort Good Hope will require money, but even so the community should not have to transition from fossil fuels immediately he says. "The five hundred people who live here should have a grace period of ten to twenty years before we have to stop burning gas." Because in all actuality that's about how long it's going to take to bring in that technology.

In Norman Wells, Imperial Oil and Canada have grown

rich from pumping oil for a century. "That's the only reason they entered into treaty with us in 1921, which was a facade and on Canada's part was not a good-faith discussion or negotiation." The government's position has always been to take full control and benefit off that oil field and try to gain legal certainty to secure their position so that we have no control explains Daniel.

Getting back that control will take Indigenous leaders banding together to exercise their rights. The capitalization off of resource extraction by corporate criminals in partnership with the government has got to end but it's going to take strong, unwavering resistance. We need to take the power back as a collective and Don Balsillie can help make this happen. Don is my former boss and the chief negotiator for the Akaticho First Nation Treaty 8 Tribal Corporation. His hair is always perfectly slicked back and he has a mob boss vibe when he walks into a room with his Elderly counterparts behind him—each one representing the communities that are a part of the modern treaty negotiation process. This is where I think that Don feels most comfortable—when the pressure is high.

Don grew up on the land in Deninu Kų́ę́. His father was a hunter, trapper and gatherer. Don owns and operates a lodge where tourists visit from all over the world, and he looks forward to his annual spring duck-hunting trip with his family. He carries his paperwork in a bison leather satchel with matching leather shoes. His family used to farm bison. "We've seen huge industrial developments around us," says Don of living in the community of Deninu Kų́ę́. "We've seen the use of hydro energy, diesel energy. Right now we're running off of hydro mostly and diesel generators for backup, and we see more solar energy panels going up not just by government but by individuals," he explains.

"We were always out in our traditional territory. Energy was a big thing, of course. You had power to your outboard motor, skidoos... dog teams too at the time, but we didn't use energy for that of course." He laughs at his joke then becomes serious again, explaining that times have changed drastically. The cost of living is higher than ever before and so, if there are other forms of energy that we can use to sustain us that are greener, then that is definitely something we should be looking into. "It's extremely high here for fuel and other commodities." There are also extreme cold spells where people consume even more energy to heat their homes in the frigid temperatures.

Don is multitasking as he talks to me over the phone. Ironically, he is in the middle of dealing with a malfunction in his furnace in the coldest month of the year. He has to put me on hold for a moment, then comes back to say, "My furnace went down and you think I could find a damn part for the furnace?" It's –35 degrees and there are no mechanics that can fix his problem in Deninu Kų́ę́. He has to wait two weeks. "Thank god I've got a good wood furnace."

Part of taking back control and implementing community-owned utilities is to negotiate with the power corporation. Currently, the NWT does not have an independent power production policy, but our Dene neighbors in the Yukon have successfully wrangled the government and power corporation to ensure that such a policy exists. Old Crow, Yukon, is the north's first remote fly-in community to have a fully operating solar farm that sells energy back to the power corporation. The revenue earned is put back into the community. They are now looking at how they can harvest willows to create a biomass facility to accompany the solar. Why can't the NWT do that too I wonder, and I turn to someone that I know will have the answer.

Darrell Beaulieu is the president and CEO of Denendeh Investments Incorporated and the general partner for Denendeh Investments Limited Partnership (DILP). The latter holds investments on behalf of all First Nation governments in the NWT. When I reach out to him by phone to get his thoughts on the NWT having its own independent power production policy he says you can't totally depend on solar and wind alternatives because they're not reliable. The sun doesn't always shine and the wind doesn't always blow, so you still have to have a baseload source. The mentality around incorporating green energy into a business-as-usual model is still very driven by cost and not by the need to protect the environment. The NWT's dependence on hydro and diesel may be reduced by the addition of solar and wind, but the capital cost of solar and wind infrastructure would still have to be factored in so that the overall cost of having these systems operating at the same time is not doubled.

Costs aside, Darrell agrees that it would be great if the switch to green energy was made in totality, making solar the primary baseload, but he says unfortunately that's unrealistic at this time. Yet we still rely on diesel as the main source of power and heat in the north, which is literally burning up millions of dollars a month.

An independent power production policy is something that Darrell agrees that the NWT needs. If Dene communities across the north want private-sector investments other than investments for utilities that are already in partnership with government-owned corporate entities, then an independent power producer policy needs to be in place, otherwise everything will continue to be completely government-funded. This is problematic because the federal government will provide the equipment for solar and wind but the territorial government will not turn it on if it's not in partnership with the local corporate utility.

On a smaller scale, says Darrell, communities would do well with microgrid solar and alternative energy. However, there is no current framework or funding for individual microgrids, and the average person cannot afford to purchase the equipment on their own.

If only they could afford it. Off-grid living is a way of becoming independent of reliance on the government because people would no longer have to pay a utility company to heat their home—which would mean that the power corporation would eventually be rendered obsolete and would have to dissolve, and that is not something that the heads of the company would want. They will hang on to their jobs with everything they got. The solution to the climate crisis in one sense is free and that is not good for the corporate agenda because the main purpose of a corporation is to generate revenue.

We need our Indigenous leaders in the north to put their foot down and demand more money for investment in green energy, as well as an independent power production policy so that communities can turn the tables and start generating own-source revenue and sell energy back to the power corporation. This will help residents, especially those in poverty, save money instead of putting most of their money toward heating their homes.

But because the NWT is so geographically politically diverse it is hard for Dene leaders to come to a unified decision when decisions about land are involved. The government is mostly to blame for this because they have worked to pit Indigenous nations against one another using divide and conquer tactics like forced competition for resources.

Indigenous nations are often piecemealing government funding together because they were not set up to be able to turn a profit like a corporation, instead they have charitable status. However, many Indigenous nations have economic corporate arms. All industries operate as corporations and oddly corporations are considered to have personhood. If a corporation is allowed to have personhood then so should nature. Until recently, nature was not allowed to be considered a corporation, but work is being done to change that to protect biodiversity as having a legal right under the common law.

There is a new leadership forum in the NWT made up of Indigenous leaders from across the north who are looking at bringing in more opportunities for sustainable economic growth under their economic branches. Darrell says that right now the primary source of Gross Domestic Product (GDP) in the north comes from the mining industry but a majority of that money is not staying in the north, it's going south because the owners of these mines are not northerners.

What it comes down to is that we need to realize who we are, says Darrell. We need to realize that as Dene we have power. We are 1.3 million square kilometers with a population of more than 40,000 people. We have a lot of opportunity to bring in innovative advanced technologies.

Earlier that morning before I spoke with Darrell he happened to have met with a prominent international investor who told him that a lot of companies will invest billions of dollars in other countries but not in Canada—especially not northern Canada—because of the lack of infrastructure as well as the stringent regulatory implications and our unsettled land claims. It's a whole mishmash of complexities that have to be worked through, says Darrell. I suppose what it all boils down to is finalizing modern-day treaty negotiations I say. For now the economic driving force in the NWT is in squalor, and until overlapping interests in land are sorted out economic opportunities implementing innovative sustainable technologies on our own terms will continue to pass us by.

Darrell would like to have all Indigenous leaders and the GNWT band together to create a strategic economic vision and take it to the federal government to demand the one thing that's going to move the needle—a socially responsible economy. To have this accomplished by Canada's 2050 net-zero target poses a challenge, especially when the attention of our Indigenous leaders is often needed elsewhere. Our chiefs are fighting a lot of different fires and are constantly being pulled in many different directions decreasing their capacity to give their full attention to climate advocacy.

We are only just beginning to exercise our collective strength again and we haven't even started to flex our muscles. We need to be more than just successful at getting investors on board, we need to dismantle the capitalist regime. If we don't then the system will be dismantled by default when society crumbles. That is what Daniel T'seleie believes.

We need to build better institutions, he says. "We need to build economic and political systems that are not founded on the same values that didn't work the first time. If we dismantle the current-state legal framework but then we start a new legal framework that has the same values then we're going to have the same problems." It's not about the individual, it's about the system that the individual is functioning in. As Indigenous people, if we get control over our own land that's good—but if we build up structures that are just a copy-paste of the colonial framework then we are going to run into the same problems that have caused this quagmire in the first place.

Daniel says that colonialism and capitalism go hand in hand and are the root causes of climate change. The mission to make more money is "so baked into every aspect of law and economic structure and life that even if one of the richest CEOs in the world says, 'I don't like this anymore,' they don't have the power to stop it." I picture a giant Frankenstein-like institution with a life of its own or a snowball rolling down a hill getting bigger and bigger until it becomes an avalanche.

The same goes for the Land Back movement, says Daniel. Even if a bureaucrat or group of judges or politicians take a stand and say that Canada is on stolen land and it should be given back, the racist doctrines that have underpinned the theft of Indigenous land are so embedded in Canadian law that it's impossible for one person or a small group of people to undo it.

The idea that one can profit off of, dehumanize and steal Indigenous land has been so aggressively pushed for hundreds of years that it is now foundational to legal, political and economic systems, says Daniel—and every facet of those systems has been invested in heavily, so undoing them would mean overhauling or eliminating the entire governing political-legal framework of Canada. This is how colonialism and capitalism, as aggressive power structures, take on a life of their own. So even if a person is very powerful and influential, they are not as powerful as the system they benefit from.

"The problem is so deeply entrenched that it requires a systemic overhaul of entire societies and nation-states in terms of how they function." To make matters worse, individual efforts to slow the trajectory of global warming are not enough. We can recycle. We can drive hybrid vehicles. Consumers can shop local. Yet this is not enough to stop the world from heating up if the world's largest carbon emitters don't step up too. It is up to the major industries to make the changes required to give us the ability to choose differently, but right now the only option we have is to buy what's being sold to us he says. Yes, tackling man-made climate change is about riding a bicycle to work, recycling and using all natural products, but it is more than that. It's about educating ourselves on what is really happening. It's about getting to the root cause of global warming.

The number one cause of man-made climate change is man's insatiable appetite for more. At the core of that hunger is money, and the default is competition and greed. As consumers we can make small incremental changes, but if media and corporate conglomerates continue to sell us overpackaged products, false advertising, misleading information, increasing global trade, then it makes it difficult if not impossible for us to stop explains Daniel. We need better options and we need a less-is-more mentality, or we won't see significant change. There are a lot of solutions to the climate crisis that have been proposed and all of them have their own complications but it's not easy to flip the switch and it's only very recently that there has been a genuine effort to try to transition away from fossil fuels.

There is an obligation on industry and government to fund the work that needs to happen for the world to transition away from fossil fuels but the powers that be will dig their feet in the dirt and hang on tooth and nail if it affects their bottom dollar, even if the consequences of doing so are dire. It's a radical concept but I believe that eradicating money is the key to solving the climate crisis and desperate times call for desperate measures. Eliminating money would drive a wedge in the corporate agenda. Just take a moment and think about what the world would look like without money. Would the world be better off? Or worse? It's hard to imagine isn't it. Almost impossible and that's because giving up money is unfathomable for most people even. Money talks.

Daniel goes on to explain that "there are potentially millions of people who will die this century because of climate change." As the world continues to warm, we will see an even starker divide between rich and poor, where the poor will become climate refugees and homelessness will be an ever-present and growing problem.

Daniel believes that the fossil fuel corporations and related interests have been working very hard for decades to prevent any genuine progress on stopping man-made climate change.

It's quite possible that they have even gone a step further to protect their interests and profits in the future by spreading propaganda to ensure that the world will continue to be dependent of fossil fuels. "Sort of like the tobacco industry," I surmise. "Yes," says Daniel. "The fossil fuel companies in the nineties hired the same spin doctor scientists that tobacco used to say tobacco didn't cause cancer, to try and argue that burning fossil fuels doesn't cause global warming." We have both obviously read the same book, *Merchants of Doubt* which addresses this concern.

Climate change is indisputable, no matter which way you look at it. The climate has always shifted, "but I think the change we are seeing now is of a very different magnitude and carries very different risks, much more severe risks, and we are potentially on track for the massive shift. What we're seeing now is fundamentally different from the natural variations of climate that we've seen in the past and that is a very serious problem." Daniel maintains that the current change in the climate is predominately due to fossil fuels, and any conversation about climate change needs to acknowledge that global extraction of oil needs to stop completely. It's not the only thing we can do, but it's absolutely necessary. "Any other approach that purports to be able to mitigate or stop climate change that doesn't also require an end to fossil fuel extraction is not going to work."

The other approach that is not going to work in the long run is hydroelectricity. Darrell Beaulieu tells me that the GNWT has been in discussions regarding running an electrical transmission line around or under Tucho and is conducting studies on its feasibility.

I scratch my head and wonder why another dam is being considered. What if the transmission line gets the green light and all of a sudden a new technology comes online that replaces hydroelectricity? That would be a lot of money wasted. We need to be thinking ten steps ahead not behind.

Plus, how many dams do we need? There are already three hydroelectric power stations impacting our northern waterways. One on the Taltson River, and two major dams upstream, the Bennett and a highly publicized controversial dam called Site C. Site C has a $16 billion price tag and a surface area of close to 10,000 hectares. It is sure to destroy riparian wetland habitats that act as carbon sinks that are home to many already vulnerable species.

Wetlands are not the only areas being impacted, sometimes entire communities get flooded out. Long before the town of Rocher River was established in the 1920s there were many Dene people living in that area. The settlers and the Dene lived as neighbors in Rocher River until a dam was proposed. The Chief at the time, Chief Snuff to be exact, was opposed to the construction of the Taltson Dam for fear of flooding, as the community was located on the banks of the Taltson River but the dam was built anyway.

It turns out the Chief was right. Rocher River flooded in the 1960s when the Taltson Dam released water. By then no one was living in the area thankfully, but many graves of the Dene people were disrupted.

Each dam upstream is connected to the Slave River, which joins to the Dehcho. With Alberta already in a severe drought, the dams will just further perpetuate the problem. When the water downstream dries up, the dams themselves will dry up—and then there will be no choice but to find an alternative to hydroelectricity once it's too late. We have been conditioned to believe that oil and gas and hydroelectricity are the only choices we have, but there are other options that are a win-win in that they are not at the expense of Mother Earth and are essentially free for us.

Right now, and into the foreseeable future, hydro is the baseload power source for the NWT. Hydroelectricity is considered "green" by industry standards but dams are partly responsible for the barges not being able to get down the Dehcho—in fact they couldn't even get out of the Hay River. The town of Hay River has begun dredging the channel in the river to make room for the barges, but Darrell questions if it's even worth it, because even if they can get out of the Hay River the barges may not make it down the Dehcho.

It is time to pull the trigger on the enforcement mechanisms outlined in the Transboundary Water Agreement. The threshold has been breached. The levy has to break. Gas was five dollars a liter in Norman Wells the last time I checked, because the barges can no longer bring in supplies which means fuel has to be flown in, making it more expensive than it already is.

Do we need Yamǫ̀ǫzha to return and help us stop the construction of these dams? Is history repeating itself? The three man-made dams upstream of the Dehcho seem to be replicating the three giant beaver pelts located on the shores of the Dehcho. The Yamǫ̀ǫzha and Yamǫ́rıa story has prophesied our predicament. Could it be that the Tsàcho has returned?

— 9 —

Firekeepers

At first, he started out filling up the water buckets. "You got to learn the basics," he says. A typical shift at work for my son depends on the day. "The beginning of the season is when we start preparing the gear for forest fires." At twenty-one years old, Trenton has worked his way up the ranks to crew leader. After only three summers of fighting fires, he has proved that he has what it takes to lead, which makes me proud.

Each day at work, Trenton does his morning briefing, he checks the weather updates from the towers across the NWT, and the indices of the fire that he is working on and how it is predicted to move and spread, then he reports back to his crew on the plan for the day. The crew works near the helipad, remaining on standby until they are sent out. The equipment they carry with them out in the field is not light. Each crew member carries a bag that contains four 100-feet lengths of hose. They also each have a pump, a pump kit and a chainsaw kit. "Once we get to the location, we set up the pump and cut a space for a helipad." If they can't find an area to land, which is often the case because there are a lot of rocks, they have to hover-exit out of the helicopter—which means they have to jump out into a few feet of water from about two meters high while the helicopter is still hovering. "You have to trust your gut instincts and know what you're doing when you're dropping into the fire," Trenton says.

Trenton sees lots of fire tornadoes when flying in the helicopter from above. "That's when you know that you can't land at that fire because it's variable winds." Fire tornadoes are unpredictable and they can reach up to fifty meters high. If a fire is big enough, it'll create its own weather. It can also block out the sun. I remember the year we thought there was an apocalypse, when the sky turned dark in a matter of minutes in the middle of the day. There was red lightning in the distance, and falling ash left a mark on everything it touched. When it rained, it rained soot.

Once the helicopter lands safely and the helipad is cut the rest of the crew members are brought out on the second trip, that way it's easier to get the equipment out says Trenton. "If the fire's small enough, we'll be able to put it out in a day and go home, but most cases we sleep out there a couple nights on the land." Each crew member has their own single tent and cot. At night, when the crew sleeps so does the fire; it smolders.

Trenton loves being in nature. Sleeping outside on the land under the stars is peaceful. He sleeps soundly even when he is only a few miles away from an active fire, it doesn't bother him. He likes being away from the noise of the city and enjoys the company of his crew who have become lifelong friends, like brothers even.

The summer of 2023 was the worst fire season on record in the north thus far. The smoke was so bad that they had to evacuate the entire city of Yellowknife. My entire family, except for Trenton, had to pack up and take only what they could carry on their back and leave immediately once the order was put in place to evacuate. Those who couldn't leave on their own accord flew out in a large military plane. My daughter in law was five months pregnant at the time and had to wear a mask as she drove down the only road out of town in a thick haze of smoke with fire raging on both sides of the highway. She was in bumper-to-bumper traffic as thousands of vehicles inched their way south to the nearest city of Edmonton, sixteen hours away. At one point she accidentally drove off the road and into the ditch and had to get help to get out. Thankfully she was able to reach her destination, exhausted but safe. The few gas stations on the otherwise lonely stretch of highway ran out of fuel, unable to restock because of the risk that the fire would reach it.

Trenton says the fire crew didn't sleep out on the land that year like they normally would because they weren't fighting a small fire, or what they call an "export fire." They were fighting an unstoppable beast. His crew and many other crews were working around the clock to put out the 100,000-hectare "campaign fire" with many other firefighters joining in to help, some from as far away as South Africa.

A fire can jump a highway with the help of the wind. It's always trying to get more air, so it keeps moving. "It's like a vacuum," Trenton says as he explains the science of fire to me in simple terms so that I can understand. He is a fire expert now and I am taken aback by how much he has come to know about the nature of fire—as well as how brave he is. "If you're really in the thick of the smoke it can be hard to breathe, even in a vehicle." He recalls being in one exceptionally precarious situation in his work truck with his entire crew, where he had to make a quick decision on whether or not to go on any further because there was no visibility. He couldn't even see the painted lines on the road in the daylight so he turned back around instead of risking it. "It's all about decisions," he tells me. You have to decide whether or not it's safe to land in the chopper, when to back out, and if you think the fire might go in a different direction. What his job amounts to is making life-and-death decisions every minute of the day.

That same summer, one of the fires that he and his crew were working to put out did a complete turnaround and headed in their direction. Through the trees the smoke billowed in like large rolling clouds. "It was a fifteen-minute hike out of the bush, we had to leave all our gear and run to the road." A mother's worst nightmare.

There were discussions at government level weeks before the evacuation that there would be strong gusts of wind pushing toward Yellowknife after the community of Behchokǫ̀ was evacuated. That same fire became so strong that Trenton describes it as one giant wall headed towards Yellowknife. That year, 3.4 million hectares burned in total from the 300-plus fires that blazed through the NWT. It wasn't just Yellowknife that was forced to evacuate. Twelve other communities in total had to leave; one town burned to the ground and one firefighter died.

Some northerners did consider staying and going out to their cabins, but Trenton warns against doing so, saying that there's fires all around and you don't want to risk going on the land. If the fire burns around you, the smoke could get really bad and you might be left with no food source. It's hard to tell this to a seasoned hunter whose first thought would be to go to their cabin to protect it or get out on their boat and wait for the fire to pass.

Firefighting is not for everyone. A physical exam is required to become a firefighter. Post-traumatic stress disorder (PTSD) and significant lung damage are both risks of the job. It's important that a person is fit to do the work, both mentally and physically. There are some people that will get out there and think they can handle the heat but next thing you know they quit right on the spot says Trenton. Not Trenton though he loves what he does. That's why he chose to stay behind to help put the fire out. He was one of about a thousand people in the sudden ghost town, most of them first responders.

When fighting a fire head-on, the crew wears yellow coveralls made of a breathable material, a hard hat and steel-toe boots. It's inevitable that their feet get wet even in waterproof boots because they are spraying down at their feet where the ash is. There's a term for this Trenton explains, "It's called blacklining."

The crews fight fire with a technique called a "backburn." Our ancestors used to practice this technique to protect hunting grounds or places of interest like sacred areas. A backburn is basically a controlled burn in the path of a fire to burn off the trees before the fire reaches them so that it runs out of fuel and slows down.

For bigger fires, crews will set up a pump, run lines, and start at the back of the fire and work their way to the front. Sometimes they split up and go on either side of the head of the fire. I did not know that there was a head and a tail when it comes to fires and when the fire is described that way it sounds like it has a life of its own and in a way it does.

The crew works on a buddy system. Trenton, being the crew leader, carries a satellite phone and GPS system with him at all times. There are always four firefighters on one crew. One crew member will use the chainsaw to cut trails through the bush so they can lay out the hose line. Typically they don't dig trenches because they wouldn't be able to dig in a straight line as the shield of rock that covers the landscape makes it impossible. Besides, it's not necessary being that there is usually a water source nearby, the north is full of water bodies big and small even though the water is receding at alarming rates. Trenton has come across bodies of water that were completely dried up while out in the field. When the world is on fire, "the land just sucks up the water. It's thirsty." Says Trenton.

It may not be scientifically proven or even researched yet but Trenton and other knowledge carriers believe that it's not just fire crews out there working to put out the fires. The land is working hard too. Mother Earth pulls water from other sources and pushes it through groundwater passageways to get to the fire to stop it. This is how the land works to save itself, and it is another reason why protecting groundwater is so important.

It's also important to try to protect the animals of the forest from the threat of an oncoming fire but when a big fire moves and spreads over a huge area in a short amount of time, sometimes the animals can't get out fast enough. Trenton has seen things he cannot unsee. Animals come running out of the bushes en masse when the fire closes in. He's seen dozens of rabbits crossing the road fleeing for their lives, bears with no fur, still alive and suffering from burns and needing to be put out of their misery. He vividly recalls a ptarmigan and her nestlings perishing. He has seen a pitiful coyote walking across hot coals with burnt paws. We tend to think only about the human loss, the destruction of infrastructure, the need to rebuild, but we often don't consider the incredible loss of habitat and how the animals have no choice but to come closer to human activity to survive—that is if they are able to make it out. Their homes are being destroyed at alarming rates. Yet predatory animals are often killed when they come close to town.

While getting ready to walk my grandson in his stroller in Yellowknife, I once accidentally put his shoes on the wrong feet and remembered what my grandma used to say: "If you have your shoes on the wrong feet you might meet a bear." And indeed a bear did come wandering into the neighborhood while we were on our walk. I had my grandson in the stroller as we passed by a steep ravine leading down to the lake, and suddenly had a feeling that a bear might be down there in the bushes. Sure enough a bear had wandered into town and made its way up that hill shortly after we had gotten home from our walk. It had most likely lost its habitat due to the extreme wildfires in the area that summer, and unfortunately it had to be put down because bears that come that close to human activity and consume human food, they will get a taste for it and keep coming back, posing a danger not only to humans but to themselves.

Then there are the rare species that thrive in the fires. A specific type of wasp to be exact. A giant *ichneumon* or more commonly referred to as a stump stabber. Trenton has a nickname for them I'd rather not repeat. He has never seen them anywhere else except for when fighting a fire. "They're about three times the size of a hornet. They lay their eggs in the burnt stump and then the heat will hatch their eggs." They don't bug you unless you bug them. Just another hazard of the job he shrugs. The wasps will die over the long winter, but their larvae will overwinter inside the shelter of the tree stump and hatch in the spring.

Even in the winter it's possible for forest fires to survive. They can burn under the snow across the north for the duration of even the coldest of winters. Winter is when Trenton is off work but he can still be found out in the bush hauling wood. He says the best trees are the trees that have been burned. It's remarkable that Trenton is not bothered by being exposed to trees. He suffered from allergies most of his life and finally underwent an allergy test. The results were that he was allergic to every kind of tree you can imagine. But he must have built up an immunity because he is not fazed by it.

Since starting his career in firefighting, Trenton sees the fires getting worse and worse every year and he doesn't know if they'll ever slow down. Eventually there's not going to be much to burn around the communities, says Trenton. "There's going to be barriers around them now." Fire crews have done the work of fire smarting and will continue to do so. Fire smarting is doing the work of making sure that communities are prepared for potential fires and taking proactive measures to protect themselves and their homes. The fire perimeter that was created around Yellowknife after the worst forest fire year on record has caused the landscape on the outskirts of the city to look drastically different.

Even after all the fire prevention work that was done, Trenton still thinks there could be more prevention work done, and the hope is that the government will see the need to keep seasonal firefighters employed all year long to study fires and bring Indigenous cultural burning practices into the forefront.

When out on the line Trenton is away from home for weeks at a time and says he misses his young family when he's out in the bush, but his job supports them. It's not like the day when a person would have no choice but to involuntarily fight fires without pay.

Back in the 1960s and 70s, the RCMP would recruit community members to fight forest fires. The two choices they had were fight fire or go to jail, Trenton's friend and fellow crew member, David Sangris, explains. Like Trenton, David is a young father. David tells me that growing up he always wanted to be a firefighter. "It's like a family." You get to travel to different parts of the territory. Most of the community firefighters in the NWT are Indigenous and work to protect their homes so there is a lot of camaraderie among them.

David says that his uncle, Chief Fred, told him that because the inside of the earth is warming up due to climate change, the whole world could burn. "That's why we've got to fight the fires." To save the world.

"You can't trust fire, it can change direction on you at any minute," David says. For him the most alarming part of fighting fires is seeing just how fast the fire can move. "Fire just rushes through the trees, it's crazy." You get one big gust of wind and a whole two hectares can burn in less than an hour. To put things in perspective, one hectare is the equivalent of two football fields.

David has been told stories from other crews of seeing animals on fire but fortunately for him he has not seen such travesty with his own eyes. He once heard about a burning moose that was suffering. "They had no choice but to add brush and burn it." There is no part of the animal that can be kept when it is burned—not even the hide, because it is singed David explains. If things got really bad, David is one of the people who would go out on the land in the opposite direction of the fire. Not many people know how to live on the land these days, he says, but he would know what to do because he's been going out on the land since he was very young. "That's why I love firefighting so much." Because for him it's also about being out on the land. "All I need is a tent and a sleeping bag and food for a couple nights."

In the morning when the sun shines right through the tent it can be overwhelmingly hot says David, but he just jumps in the lake and washes his face. Then either he or another crew member makes breakfast, afterwards someone washes dishes, and they get to work—which he calls "boots on the ground" or "staying in the burn." The crew walks down the line following the burnt trees. How far they walk depends on the distance to the fire. They might walk one kilometer or five kilometers. It's a lot of walking, which makes for long twelve-hour days. When they get back to camp, they eat and go straight to sleep exhausted. "Some of us are so tired we just go straight to bed without eating."

Some guys can't handle it, says David. "They'll go to a fire once and get home and quit. It is really, really hard work, but at the end of the day it pays off. You feel good doing all that work." There is great purpose and sense of satisfaction in the work they do—after all, they are in the business of saving lives.

Trenton and David are just two of the many firefighters across the NWT who work on the front lines. They have a deeply personal understanding of what it's like to battle the raging flames. Their entire summers are spent fighting fires and the summers are what most northerners look forward to but that's the peak of the fire season so firefighters have no downtime to enjoy the short northern summers, but it's a sacrifice they are willing to make.

Trenton and his crew fight both the heat of the fire and the heat of the hot summer sun beating down on them all day long. I don't know how he can take on such grueling work day after day and still love it at the same time. "Don't you get really hot out there being next to a forest fire and under the blistering hot sun?" I ask but he just nods and says "Sometimes."

Most of us will never know what it's like to have to fight fires in a heatwave, or witness up close how a fire generates its own storm system, or bat away invincible insects that thrive in flaming infernos. Community firefighters deserve the highest honor.

Cassandra Blondin Burt is the great-granddaughter of the acclaimed Dene writer George Blondin, who wrote about Dene medicine power in his books *Yamoria the Lawmaker, When the World Was New* and *Trail of the Spirit*. In his books he writes about the spirit of fire. Fire is often a central feature in Dene stories, we even have a man named No Fire who appears in the *Book of Dene*. We also have a story of why the fox's tail is black on it. One time Fox had used his tail to set fire to Raven's teepee to free the caribou because Raven was hoarding them all to himself.

In the story of *Yamǫ̀ǫzha and his Beaver Wife,* Yamǫ̀ǫzha cooked the giant beavers over a fire and the grease that drizzled onto the flames began to burn—creating a plume of smoke so big that it rose to the sky and can still be seen to this day in the Dehcho.

We have been conditioned by colonialism to view fire as something that is negative Cassandra says in frustration. "Wildfires are like a fever in the body." The fever is there for a reason, says Cassandra. The fever needs to burn through the disease. The disease of destructive human presence. If we create a safe space for someone who is in a deep fever, it helps the body come back to health. This applies to Mother Earth too.

Cassandra is well informed on Indigenous fire practices being that she was contracted by the federal government to conduct research on the subject.

Cultural burning or what Cassandra calls fire medicine "is not about setting fire to everything. It's about a cool burn. It's about very intentional practices. It's about a real, close and intimate relationship with the land." Cassandra explains. "Fire is actually one of the blessings of the land." It has a purpose. Fire is Mother Earth's way of cleansing she says. Forest fires replenish the forest floor and makes way for new growth.

The land has its own methods, its own medicine, to transform and break free from disease. "We've actually created a lot of disease by preventing fire. We could intentionally work with fire to support what nature is already doing, or we can get the heck out of the way, but us trying to control fire is an illusion of Western control," says Cassandra.

The government needs to loosen the need to control and suppress wildfires by creating a safe space for fire to burn. Engaging communities on how to best incorporate cultural burning practices is the best place to start. Working with Indigenous knowledge carriers is important, because it is becoming top priority to protect, not only people's homes, but the land so that we don't lose our boreal forests.

Steven Nitah explains that if we are just going to let fires burn out of control, it cancels out biodiversity efforts because the amount of carbon dioxide that is released into the atmosphere from a forest fire is massively contributing to the greenhouse we now live in. If we are serious about reducing carbon emissions to meet world climate targets, then we need to place priority in managing fires even if that means it's coming out of taxpayer dollars because fires are only forecasted to get exponentially worse.

Most insurance companies have categories of prescribed burning for insurance holders and are now looking at creating a new cultural burning policy with a growing interest in Indigenous-led fire stewardship that may help with their significant losses due to the frequency of forest fires over the years. This would lower the high cost of customer premiums that insurance companies are now having to charge in high-risk areas.

Fire has been humanity's mode of survival since the very beginning of time. It's the one thing that separates us from animals. Fire is central to our livelihood in the north. It keeps us warm during the long winter months. When you put wood in the fire it is very important to use gloves to protect your hands, says Mary Jane Cazon. A tree is very powerful. Trees have a spirit. "When you just use your bare hands, and you throw wood in the fire your hands tend to get wrinkled fast." It might sound like an old wives' tale, but it makes sense. As we age our skin begins to resemble a piece of wood, with deep lines marking our years like the rings on an old tree stump.

I forgot about this teaching when adding logs into the woodstove while spending three nights in a cabin alone in the Yukon last winter. If I want to one day live out in the bush I need to get used to cabin life. The cabin I was staying in was stocked with enough wood to last me the whole three days, all chopped and neatly piled right outside my door. The first night in the cabin I woke up freezing. The fire had gone out in the middle of the night because I didn't tend to it. The second night in the cabin I thought I was smart and stuffed the woodstove full of logs so that I wouldn't have to get up in the middle of the night to put more wood in but boy was I wrong. I woke up in the middle of the night drenched in sweat. It had to have been plus forty in the tiny log cabin. It was so hot I couldn't sleep. I had to open the cabin door to let the cold air in. The last night in the cabin I found the middle ground. I woke up every few hours in the night to stoke the fire. Now if I could only get good at chopping my own wood.

We feed the fire for warmth, but the Dene also feed the fire for another reason: to honor our ancestors who have gone on before us. We put tobacco or even food into the fire as we stand around in a circle and say a prayer. The feeding of the fire is a sacred ceremony. Through this practice, we also give thanks to and honor Creator.

We also have firekeepers or fire carriers whose role is to keep a fire going during important gatherings. The role of a firekeeper is only given to certain members in the community who have shown that they can carry the torch. During a Dene Nation gathering at West Point First Nation, in the South Slave Region, the Chief of the community tasked my son and his friend with carrying that responsibility and they gladly accepted. For three days Trenton and his friend slept inside a teepee next to the band office. They stirred the fire in the middle of the night and added more logs when needed taking turns to sleep to make sure that the fire didn't go out. They did this for three nights until the decision was made and a new Dene National Chief was elected.

In Jonas Noel's small housing unit in the old folks' home in the center of Ndılǫ, the wind blows a gust of snow into his kitchen whenever he opens the front door but it is not cold inside because he has the heat turned up high. When I ask him his thoughts on the forest fires Jonas says that they are getting more dangerous every summer. There will be no more food for the animals to eat because of the fires, and animals will slowly disappear. The most dangerous element on Mother Earth is forest fire, says Jonas. "There will be end of the world someday with the forest fire." Jonas's prediction is a warning that we must stop adding fuel to the fire.

Gilbert Cazon is still hopeful and shares a message with me that one of his Elders told him which is that the land is going to go through a cycle of hot and dry conditions. There will be a time when all water will evaporate, and because of that the land will burn, "everything is going to burn all at once." Once that happens it will create charcoal. When the rain comes, the charcoal will filter out the toxic substances including carbon dioxide in the air, land and water. "The land will take care of itself."

Gilbert is onto something. The fancy term for the byproduct of a forest fire is called biochar and it can be used to trap carbon according to researchers. This is a fairly new advancement in the search for climate solutions but it's promising and not the least bit surprising to the Elders who have known this all along. I think about my grandma and how she inherently knew that charcoal was a cleansing agent. If I had a cut she would tell me to rub ashes in it. My grandma would say that the ashes from a fire are healing and in that way, fire serves a very important purpose. Yes, fire has the potential to destroy the world like Jonas says, but if we are wise enough to use fire to our advantage and see it as a solution instead of a battle that we must conquer, it could potentially be what saves us in the end.

If we think about fire as an emotion then fire is most definitely anger because it is often associated with feelings of fury, rage, temper. On the contrary, rain is often associated with feelings of sorrow, wind is associated with change, sunny days are associated with happiness. Trying to control a wildfire is like telling a person to calm down when they are upset, it doesn't usually go over too well. Mother Earth needs an outlet for her anger, she can no longer suppress it. Anger is usually a good indication that something needs to change. Not allowing Mother Earth to express her emotions will cause further outbursts of wildfires. This is why we need to practice controlled burns. Anger is a symptom of a condition that is unbearable. Mother Earth can no longer bear the weight we have offloaded onto her to carry. We have made the conditions for her intolerable.

We need to stop thinking about anger as a negative emotion. Anger is a defense mechanism. It is the most powerful way to get a point across. Look at it from a mother's perspective. Mother Earth has had a lot of patience with us, her children. It has taken a lot to put her over the edge but she's reached her breaking point and she is now putting her foot down. We can no longer get away with not listening to Mother Earth's warnings because the wildfires raging across the planet won't let us. Mother Earth is using fire as a disciplinary action so that we have no choice but to go on a time out.

— 10 —

Sovereignty Is Not for Sale

I made the mistake of romanticizing negotiations before I ever sat at a negotiation table. The word *negotiator* has been glorified in movies where mostly men in black suits have mastered the skill of talking someone out of blowing up a building by convincing them to take their finger off the detonator. In all actuality the process of negotiations is quite boring in my experience. There is no one standing on a ledge with police surrounding the building trying to talk someone out of jumping. There are no hostages—unless you consider the fact that I felt held hostage for eight hours a day, stuck in a fluorescent-lit air-conditioned room with bad coffee and stale croissants, having to make small talk with the federal government's lead negotiator about her prize show horses. The type of negotiations I found myself were mostly just a lot of technical wordsmithing among lawyers getting hung up on simple words like *may* and *must* which dragged on for hours, days even.

There were times that the two sides of the table shared heartfelt discussions one-on-one at the back of the room about how it was difficult to be away from our families each month to travel to the negotiation meetings. It surprised me how personable everyone was with one another even when representing different interests. There were three parties at the table—the Akaitcho Dene First Nations, the federal government, and the GNWT—and every so often laughter kept the mood light. Everyone knew each other and seemed like old friends, but when it came down to discussing land interests the tone would turn serious.

Akaitcho negotiations have gone on for over twenty-five years now, and there is still no end in sight. It is not uncommon for negotiations of this magnitude, with hundreds of millions of dollars on the table, to take many years to reach an agreement. With revolving governments and shifting mandates, negotiations can easily come to a complete standstill. Government negotiators arrive at the table with strict talking points given to them by their superiors, whose instructions are that they are not to agree to anything more than a certain number. Even though the federal government prides itself on its mandate to advance reconciliation, they are still not budging when it comes to land as dually noted.

When hearing the word *reconciliation* I think of what Dene Elder Francois Paulette once told me, "Reconciliation is meaningless unless we are reconciling with Mother Earth and Indigenous people together. This is done by undoing and restoring the damage that has been done to Mother Earth. Man can't continue to delude themselves of what harm they are doing to the future of their children."

Francois is a founding member of the Indian Brotherhood of the NWT, who I had the opportunity to interview when writing a piece on climate change for the *Indigenous Peoples*

Atlas of Canada and he does not mince his words, reconciliation is still a pipe dream.

> "Man can't continue to delude themselves of what harm they are doing to the future of their children."

In the case of the Akaitcho agreement, the amount of surface and subsurface land rights to be divvied out is under confidential discretion. In the case of most modern treaty negotiations, the government is posturing by giving few choices, either agree to less surface rights that include an equal amount of subsurface rights, or agree to having more surface rights precluding any subsurface rights at all. Essentially, having no subsurface rights means that anything above ground would belong to the Indigenous government but anything below ground would be off the table. The nation without subsurface can't so much as put a shovel in the ground. It's like picking the lesser of two evils. It's a game that our side of the table shouldn't have to play because we have had so much taken from us already.

One of the main points of contention in modern treaty negotiation processes is that the federal government and Indigenous people view the land very differently. It is the same problem that caused contention in the first place with the signing of the original treaties over a century ago. The same empty promise of Indigenous rights and title, is just being repackaged and presented by the government with a pretty bow and a hefty price tag. There are existing precedents that have set the bar high for Aboriginal title. Landmark cases like *Calder, Sparrow, Delgamuukw* and *Tsilhqot'in* where the courts have recognized and acknowledging Aboriginal title but even with these wins for Indigenous people, the government will not let go so easily, and the burden of proof still rests on Indigenous peoples to prove their inherent Aboriginal rights and title.

Many First Nations leaders are sitting at the negotiation table right now all over Canada working on strengthening the modern treaty agreements, but if land rights and title are not included in the proposed self-governance regimes, then true sovereignty will not be achieved. True sovereignty will be when the government no longer infringes on Indigenous lands.

Most modern-day treaty agreements include provisions for public health, conservation and acts of war as an excuse for the government to infringe at any time. Climate change response could easily fall under any one of the above categories. The government consistently operates in a very patronizing manner towards Indigenous people. Most government officials don't believe that we are capable of governing ourselves and implementing and enforcing our own laws. There is a fear from the general public and even Indigenous peoples as well that we will not govern those that live on our land competently and this stems from witnessing the prevalent trauma that Indigenous nations have endured but are beginning to heal from. Even so there is often distrust in band elected leaders because of a few bad apples on Chief and Council. Still, we march on steadily to rise to the occasion of asserting our rights to be sovereign nations. We are capable of being sovereign nations without government buy-in. Ultimately, I don't see what the point is in negotiating modern treaty agreements at all when we can override the government's game and bypass having to play it at all. If Indigenous people do not have the ultimate say over what happens on our land, or are not able to govern ourselves accordingly as sovereign nations equal to the Canadian state in these negotiation rooms then we are within our right to walk away and rely on the original intent perceived by the Indigenous people at the time of signing the original treaties.

The Dene have never stopped exercising their sovereignty. Dene leaders of the past died fighting for their rights. It would not be right to settle for the crumbs that the government is bringing to the table. We must ensure that our ancestors did not live and die in vain. The sale of the land is not something that was accepted then and it won't be accepted now. The land is simply not for sale.

There are three self-governing nations in Dënéndeh: the Tłı̨chǫ, the Gwich'in in the Iñuuvik Region, and the Sahtú communities around the Great Bear Lake. Akaitcho and many other nations are currently at the negotiation table. Akaitcho is the umbrella agreement for Deninu Kų́ę́, Ndılǫ, T'èʔehdaà and Łutsël K'é as part of Treaty 8, which was signed in 1899.

The name Akaitcho comes from the leader of the Dëne Sųłıné who was alive back in the late 1700s. His name, Ekècho, translates to "big foot," because he was so tall.

Perhaps Ekècho's most notable contribution in the history books was that he led Sir John Franklin and his men on their Arctic expedition. The main street of Yellowknife is named after Franklin and not Ekècho, and there has been calls for a long-overdue name change that would honor the life of Ekècho rather than giving credit to the explorer from the other side of the world.

Don Balsillie, the chief negotiator of the Akaitcho agreement, has worked long hours over the years leading up to the establishment of a draft agreement-in-principle, and he has had to update new chiefs and councils many times over the years on where the agreement is at because the First Nations band election process in Canada operates within a government structure where there is an election every four years which makes for a lot of turnover with little to no succession planning. In the days before contact, Dene leaders were trained up from a young age and served their community their entire lives, gradually building up the shared wisdom to be a part of decision-making. This no longer occurs unless it is happening strategically in hereditary bloodlines. The imposition of the Indian Act formalized and designated band electoral voting systems. Before the Indian Act, Dene families and communities would travel far and wide to gather for celebratory events or hold meetings regarding important matters, but today that's not really happening either, unless it is for a political assembly where the slow bureaucratic voting process often diverts attention away from other more urgent issues like addressing the climate crisis.

The Indian Act contributed to the marginalization of Indigenous people that still exists today. It has divided our nations and pitted us against one another.

While our leaders are distracted, non-Indigenous folks are leading and creating policies and laws on our behalf They are even learning our languages to be able to converse with Elders to get them to agree to projects. If Indigenous people are not part of the development of those policies and laws then it's being done for us, without us. To fix this problem an economic power shift needs to occur. In her book *Indigenomics,* Carol Anne Hilton writes that Canada's land base covers 998,500,00 hectares. Today, only 20 percent of this land is directly controlled by Indigenous people. This percentage of land base that is in our hands, when compared on the global scale, is substantial she writes. Indigenous peoples are "economic powerhouses" waiting in the wings.

Indigenous rights holders have a lot of power that is not being exerted—but that is starting to change. There is a new sphere of influence taking place, says Hilton. A prime example of how this power is being taken back is the establishment of Indigenous Protected Conservation Areas (IPCAs) like the Thaidene Nëné, meaning "land of the ancestors" in Dënesųłįné. The sacred area to the Dene that covers 26,376 square kilometers is now designated a national park reserve. When Łutsël K'é announced that their new national park had been approved, it was a huge celebratory milestone for the community and I was invited as part of the Akaitcho negotiations team.

I don't think I'll ever get used to flying in a bush plane. Bush planes are loud, cramped and bumpy and some of the pilots look like they are straight out of flight school. I try not to think about the story that Elder George Mandeville told me about almost being in a bush plane that almost crashed in the Mackenzie Mountain range. "We had that plane so full of gear that the pilot had to take two seats in the back out so he could fill it up." George remembers sitting on top of the gear. There was no seat and no seatbelt. He and his staking partner, Hugh Arden, were planning on camping out for a month on the land to prospect for lead and zinc. "The Mackenzie River was to our left about five miles, and to the right was the Horn Plateau." They were flying at about 10,000 feet when all of a sudden the plane stopped mid-air. "The prop stopped moving. It was just deathly quiet." The pilot radioed for help and gave coordinates.

"It seemed like we were dropping straight down," George recalls. He'd figured the plane would glide but even if they did manage to glide to the ground there was no water around for them to safely land with the skis. "The pilot panicked. He took his seatbelt off and started crawling to the back." George had to move out of the way. He thought maybe the pilot was going back there to fix something from the back of the plane. "He didn't say anything." George looked at his partner and then looked back at the pilot. "We only had so much time." It seemed to be that the pilot had given up and retreated to the back of the plane in fear. That was when George's partner grabbed the pilot by his belt and threw him back into the cockpit and yelled, "Try something. Try the starter. Keep trying. We're gonna die." The pilot did as he was told. He pressed a few buttons and the ignition worked. "It fired up and started sputtering."

The treetops were so close that they only just cleared them before picking up speed and climbing back up into the air. They were "a couple seconds from death," George says.

With multiple copies of the community newsletter in one hand, the draft agreement in principle in the other and two computers in tow, my office hung off my shoulder as I headed to the small hangar down the road from the airport in Yellowknife. Myself and the rest of the Akaitcho team would be catching the Twin Otter to go to Łutsël K'é to update the community on the status of negotiations. The plane had just enough seats for the five of us, plus a few other passengers who were flying home. To my dismay, there was only one pilot. He was middle-aged, big and tall, with red hair and a big bushy red beard. He looked more like a lumberjack than a pilot. He talked us through our safety and emergency exits and then slammed the door of the plane like it was a big rig. It didn't close properly the first time, so he had to open and shut it again to make sure it was sealed. "Should be a shorter flight today, folks, we have a tailwind," he said before starting up the sputtering engine. I held my breath and closed my eyes as we took off.

It was a good takeoff, smooth. Once my fears quieted to a whisper, I felt brave enough to look out the window. I wiped the soft, thin layer of frost off the plexiglass with my bare hand, and the imprint looked like the side profile of a woman with her hair blowing in the wind. I looked through the wild woman's outline and stared down at the ground below me. It was a cloudy day, but mild—no turbulence. As I looked down at the ground I noticed the deep lines in the rock formations. Mother Earth is aging, I thought. She has deep grooves in her surface, wounds that tell an ancient story. Don leaned over and yelled above the loud engine, breaking the barrier of my earplugs, telling me that the fault lines in the rocks are from the ice age—and the reason the cliffs had formed was because the earth had split apart, shifting like a tectonic plate. Referred to as a "sheer zone," it was once an active volcano during the Proterozoic times he says. The fracture in the earth had exposed minerals galore. A paradise for eager prospectors, which is why Łutsël K'é has had to find ways to protect the area from development—hence the establishment of the park. I nod taking in the brief geography lesson. We soon reverted back to sitting in silence and I watched as one of the community negotiators, sitting in front of me, a hunter at heart, pressing his forehead to the small window probably trying to spot a caribou or musk ox I figured.

Relieved to have landed safely, the first thing I saw was a large black dog running loose on the runway to greet us. The airport is a small house and visitors are greeted by a hand-painted welcome sign on a sheet of plywood. A few people that were waiting to catch the same plane out, lined up by the door as someone played an acoustic guitar in the corner of the room. The dog somehow got into the building and ran around excitedly, sniffing and hitting everyone with his long tail.

Our ride was already waiting for us by the time we grabbed

our luggage. It was the community negotiator for Łutsël K'é, driving a big white work van. He drove us straight to the band office, which was only five minutes away, down a winding dirt road. The floors of the office are made from local rock—big slabs of granite from the cliffs we'd flown over. A small musk ox head was mounted on the wall in the meeting room. "Same size as a caribou," an Elder said to me when he noticed me looking up at its lifeless eyes. We were served caribou stew for lunch, and a tasty fresh fish soup with soft baked bannock paired with homemade blueberry jam. Another dog stood patiently outside the door, watching us eating and hoping for scraps.

After lunch we all got a ride on skidoos to get to the community hall, which was set up for us to give a presentation on the status of the negotiations. I fiddled with the projector to try to get the presentation up on the big screen, and gave up when it kept blanking out on me, so we relied on the paper copies I'd brought. Not a complete waste of paper, as it turned out. In the corner of the hall, a bingo machine and drum kit collected dust. In the main foyer around the corner, a group of Dene men sat around watching sports and drinking endless refills of coffee.

The meeting began with a prayer from one of the Elders. There weren't very many people in attendance, mostly adults and a few youth. People came and went as they pleased. Some stopped in to see what we were there for and left again to smoke outside. Some stayed and listened intently. It would have been nice to have the high school students attend to get credits but learning about their land rights wasn't part of school curriculum.

Don, our chief negotiator, began by updating the community members on where the negotiations were at. Once he finished his spiel, the floor was open to questions. The questions mostly revolved around what governance might look like once the agreement went through. We received many reminders to always protect the land. I took mental notes.

That evening the charter plane arrived earlier than expected and we had to leave in a rush. Some community members were not pleased to see us leaving so early and suggested that we stay overnight. I agreed but wasn't in charge of logistics. If I was, we would have stayed for a few days and held workshop gatherings with the youth, knowledge carriers and Elders.

As we reversed out of the hall parking lot in the big white van in a hurry to catch our flight, I felt like a typical government worker just dropping in and out. The van sped along the snow-covered road that led up the hill to the airport, where a big raven's nest sat atop a lonely lamp post. The pilots had turned off the plane's twin engines while they waited for us. At least there were two pilots this time, I thought. We lined up and waited outside in the blistering cold until they pulled out the stepladder for us to get in.

On the way back to Yellowknife I looked out the window again at the world below. When flying over Tucho I saw large slabs of ice had slammed together so violently that it formed a large rift in the otherwise-smooth surface—creating one large solid wave above the cracks.

That trip taught me one very important lesson: that our integral responsibility as Dene people is to lead the way by bringing our teachings with us into the future for the health of Mother Earth. We must remain steadfast, reflect on our ancient principles, stories and laws in the face of adversity, and resist government and corporate demands for access to land rights that run the risk of causing further irreversible damage to Mother Earth. The ongoing impact of colonialism often puts pressure on Dene leaders to give up land rights in exchange for financial incentives and the promise of jobs. While the potential of economic stimulus, even a temporary windfall may seem promising, it ignores the truth that colonial governments created the dire financial situation that many of our Indigenous communities are in, and we just cannot stand by and allow the government to call the shots anymore. There is no room for apathy. We need to have a veto when it comes to making decisions about our lands and water. Like Ekècho, we once were warriors, and our warriors will rise again.

There is a story about war and peace among Ekècho and another Dene leader named Edzo, the leader of the Tłįchǫ, that took place over 200 years ago. There was a time when a fight broke out between them. They did not fight with sword and pen like the government, nor did they fight with fists and weapons, instead they fought with medicine power. The war between them went on for a long time until a Dëne Sųłıné man named K'àtehwhì, who was married to Edzo's wife, acted as mediator for the two warriors and brokered peace between them.

Even back then the Dene had negotiators.

I had the honor of driving with the late Elder Eddie Sikyea on a trip to Behchokǫ̀ to learn more about what we now call the peace agreement between Ekècho and Edzo from a group of youth who had been working on an animated cartoon based on the events that took place at Gots'ǫkà Tì or Mesa

Lake. Eddie was dressed in a black leather jacket, his shoes looked freshly polished, and his shirt was pressed and buttoned to the top of his collar. His black hair was held in place with a good amount of gel, and he leaned heavily on his cane outside of his apartment building as he waited for me to pick him up. The language barrier between Eddie and I hung in the air, so we sat in silence for much of the drive. When we arrived at the gym I doted on Eddie, offering food, tea and water, opening doors and pulling out his chair for him.

Eddie's father was a very respected Dene Elder as well. His name was Michel Sikyea and he was a hunter and a community leader. He was a signatory to Treaty 11, yet that did not stop the government from charging him with a crime for shooting a duck in the 1960s in the vicinity of Yellowknife. Michel had to go to court to fight for his right to hunt. At the end of the day the legal costs equaled a million dollars, while Michel was fined one dollar after he won his right to hunt in the senseless court hearing.

The "million-dollar duck" battle illustrates how Indigenous people are constantly having to fight for their inherent Aboriginal rights. The right to hunt, fish, trap and gather is seen as incompatible with some conservation strategies, which is why it is important to clarify that Aboriginal rights do exist within conservation efforts—not apart from them. IPCA's are one way of getting around this but it's important to remember that historically conservation has excluded Indigenous peoples. Eddie's father is not the only one to have to take the stand as a criminal in the eyes of the court to defend his hunting rights. There is a long line of Dene hunters who have been pulled into court.

Thankfully, we now have the United Nations Declaration on the Rights of Indigenous Peoples (UNDRIP) as a primary defense to affirm that Aboriginal rights and title exist apart from and both inside and outside of conservation efforts, but there are still upsets—like the time a group of Dëne Sųłıné hunters got in trouble for shooting a caribou. The caribou happened to belong to the Bathurst herd which was off limits for harvesting due to their decline. It was shot outside of Thaidene Nëné and brought into a camp in the park. The camp was later raided by GNWT officials.

The incident made media headlines. The question was which law superseded the other. Was the constitutional protection to hunt overridden by the territorial Endangered Species Act or vice versa? The park was supposed to safeguard Aboriginal rights to hunt but the blurred lines were whether that included endangered species or not. The situation has since been rectified but it goes to show that capturing every single possible scenario can't be done on paper which is why living documents need to be created so that

"We have a deep understanding that we are a part of nature, not above nature."

they can be adapted and amended with our changing world where paramountcy among the colonial government and Indigenous nation governance needs to be fleshed out further.

Łutsël K'é Dene First Nation has responsibility for the management of the Thaidene Nëné, in partnership with the federal government, and has the final say over all decisions regarding the park.

Steven Nitah, who has many years of experience working in IPCA's, says that Indigenous people are the experts when it comes to conservation. "We have a deep understanding that we are a part of nature, not above nature." The world of science is greatly influenced by Catholicism and Christianity, he says. "In the Bible it says you can bend nature to your will." Steven calls the Bible the first corporate book. He once stood up in a room full of scientists and told them, "My knowledge and my people's knowledge is not for sale. Science is for sale." There are scientists that will sign off on every major project that contributes to climate change today, yet the ones who speak out about climate change are crucified he says.

We need to incorporate Dene value systems into business models, Steven explains. "We are too late to change capitalism," so instead we must look at how we can incorporate Indigenous value systems into the capitalist system by adding financial value to nature. "Right now, nature has no value until you tear it up and create something out of it. If we are going to create value from nature, then let it be Indigenous values that are represented." To protect biodiversity, when we are talking about building a road to access remote resources like minerals, for example, "we might have to wind the roadway around a wet land instead of draining it." It's about going the extra mile to protect nature instead of cutting corners and destroying it.

"My knowledge and my people's knowledge is not for sale. Science is for sale."

Steven says that climate change occurred naturally in the past. This has been proven in ice, sand and rock core samples. However, the climate is changing rapidly now because of the amount of carbon that's going up in the air is like never before.

There's too much carbon in the air and there is now a global market to sequester carbon and to eliminate it as much as possible, explains Steven. "For the first time there's a value put on nature that doesn't require nature to be destroyed to create a product." For so long, nature has been viewed as having no financial value until it is destroyed and a product is created out of it. A tree can be cut down to make paper but keeping that tree intact prevents carbon from entering into the atmosphere—and that is something that is more valuable now than ever. A tree is worth so much more standing.

Steven tells me that nature reserves like IPCA's are one-third of the solution to mitigating climate change. Ecosystems naturally sequester carbon, and the healthiest natural ecosystems in the world are located on Indigenous lands because they have not been exploited for gain. Steven says that Indigenous people hold the key to making space for and maintaining a healthy relationship with nature and deserve to have that recognized and be given the platform to voice our value systems globally. I agree, but it seems to me that the doors to those rooms aren't opening fast enough—especially when many Indigenous peoples from across the world are only observers at places like Conference of the Parties (COP). But Steven gives me hope when he says there are more and more Indigenous people attending COP, not only as observers and activists but as decision makers. He also assures that the agenda of reconciliation is a form of land back through conservation, and soon large corporations will have to choose between paying a significant amount of tax or purchase a carbon credit to protect biodiversity, thus investing in IPCAS. He calls carbon credits "nature stewardship credits." One credit can be purchased for up to one hundred dollars annually and is equal to one acre of managed lands within an IPCA. Steven is one of the top advisors on how to roll out these nature stewardship credits globally. "It's not a market designed for longevity," he says, but until our carbon emissions are mitigated carbon credits are a necessary solution.

Steven's job has taken him all over the world. He was in the room when the Montreal Accord was adopted to protect 30 percent of land mass globally. Canada has committed to protecting 30 percent of its land, inland waters and oceans through IPCAs. Indigenous nations would own the IPCAs and get some semblance of land back as part of the reconciliation process. In turn, Indigenous nations can generate a revenue stream that will provide their communities with a stable economy and on a wider scale support global health explains Steven.

"It's great to have that 30 percent but that is not good enough by 2030 anymore." Canada is sitting at only half of what was promised. This leaves a lot of opportunity for Indigenous peoples to implement IPCAs within their territories. It's in this biodiversity space where reconciliation can start, says Steven. And it's important to ensure that any conservation efforts have solid protection built in against mining and development—now and into the future.

Retired lawyer Daniel T'seleie is also working to implement the Establishment Agreement for an Indigenous protected area in his home territory called Ts'udé Nilįné Tuyeta. In Dene Kədə́, Ts'udé means "big tree," Nilįné is any tributary of the Dehcho, and Tuyeta means "wetlands." The site, says Daniel, has always been a very special area in terms of resources. It is an important area for harvesting, used by many Dene people at different times of the year. Although maintaining the land and water has been happening since time immemorial, Daniel is now doing the work to ensure that protection is reflected in colonial law.

On the issue of industrial and extractive activity that might impact the agreement, Daniel explains that the protected area will have a buffering threshold to monitor outside activity. Right now the bulk of the work that he is doing is drafting a management plan, led by a board made up of Dene people from K'asho Got'ine. Like many educated Dene people living in both worlds—otherwise referred to by the Tłı̨chǫ as "strong like two people"—Daniel is well versed in both Dene epistemology and Western epistemology, and wears many hats. He is also the chief negotiator for the K'asho Got'ine self-government negotiations. The mandate he has been given by leadership is to do what he can to get back control of the land he explains. The negotiations he is leading are distinct from the Sahtú Dene and Métis Comprehensive Land Claim Agreement that was signed in 1993.

What continues to come up in his work is that the K'asho Got'ine Dene want full control over what happens to the land. Daniel expresses his frustration at the formal process of the negotiations, which limits discussions about land and title within the municipal boundary and pockets of land that exist around the Sahtú settlement area. Essentially, large swathes of Crown land are considered off the table by the GNWT and the federal government of Canada, because from their perspective those lands were already addressed in the Sahtú land claim where title is said to have been ceded and surrendered to the government. That is still a particular issue for people of K'asho Got'ine. "People don't want to give up the land," says Daniel. "If we're going to stop climate change, if we're going to stop global ecological collapse, we need the collective advancement of Indigenous sovereignty and self-determination."

Historically and right up to the present day, settler states and colonial governments all over the world have not done a good job of governing the land respectfully, and this has been an extra strain on Indigenous peoples. Many Indigenous nations have been trying to stop the irresponsible actions and policies of the government, and they are doing far more for the actual protection of biodiversity and land than any other government, entity or nation says Daniel. Which begs the question: Why are Indigenous people still being told they don't have control over their own lands, when in all actuality they have been taking care of their lands the entire time and picking up the government's slack?

Where Indigenous nations once placed trust in the Canadian government to uphold the treaties, the government now needs to trust in Indigenous people. Thankfully for everyone involved, vindictiveness is not found in the Dene Laws.

The Mackenzie Valley Land and Water Board and the Mackenzie Valley Environmental Impact Review Board (MVEIRB), has tried to ensure a government-to-government management plan. MVEIRB was initially established to implement legislative systems of land and water across the NWT ultimately, our collective sovereignty is not for sale. The concept behind MVEIRB arose out of back to the 1970s and the formation of the National Indian Brotherhood—which eventually became the Assembly of First Nations (AFN). The Dene leaders of the day had banded together to stop a proposed pipeline—which would transport oil from the Arctic Ocean through the Mackenzie Valley watershed to southern markets.

The process of decision-making within MVEIRB starts with a preliminary screening process, any new proposals that require a license, permit or other authorization must go through this stage first. Land and water boards, depending on the region the application is made in, will review and manage the application. Land and water boards are made up of First Nation delegates within their respective region. For instance, the Tlicho government has its own land and water board called Wek'èezhìi with representatives from each of their four communities.

The first stage of decision-making is usually a quick determination of whether or not an application for natural resource development will have a significant impact on the Mackenzie Valley or cause public concern. If it is decided that the proposed development would pose an issue, then the proposal moves on to the second stage, which is an environmental assessment process conducted by MVEIRB. Prior to MVEIRB's existence, no environmental assessments were taking place at all and the framework has become something to be modeled after.

In the second stage a more thorough assessment is done and a recommendation is made to the GNWT Minister of Lands on whether or not the project can proceed with subordinate regulations and licensing or whether it should be rejected altogether. Additionally, there may also be a request for a review by an independent panel. Lastly, an environmental impact review is conducted and a focused study on issues raised in the environmental assessment stage is compiled.

The requirement of consulting with Indigenous governments is triggered after the initial discovery stage. That is when section 35 of the Canadian Constitution comes in, which states that any development that is to take place on

Indigenous lands must go through extensive consultation with First Nations.

For far too long, industry and government have ignored the right for Indigenous northerners to sit at the economic table, and have failed to consult genuinely and robustly before commencing projects that have the potential to impact the land and water. Section 35 makes it so that anyone who wants to develop a project in the Mackenzie Valley must consult the Indigenous nations in their regions, both those who have settled their land claims and those who haven't.

Many Indigenous leaders across the world are taking a stand against industrial development in order to prevent environmental injustices from occurring, even when offered substantial incentive—leading the way for others to follow. There are Indigenous leaders and who have accepted poor deals and have sold their land. They may have regrets after the fact but the reality is that some Indigenous leaders have lost sight of the ancient Indigenous legal orders—the original instructions they were given that prioritize caring for the environment. It is imperative that they are guided back to standing up for their earth mother, reminding them that, the land and ultimately, our collective sovereignty is not for sale.

— 11 —

Mother Earth's Medicine

My grandma was born a twin, but her twin sister died shortly after she was born from tuberculosis. Her name was Dora. Tuberculosis is a long-lasting disease with the main symptom being a bloody cough which many Indigenous people across the north were exposed to when the illness was brought by settlers. Some people were hospitalized for years at a time, and would sometimes be experimented on by doctors without proper pain medication. Just like residential schools, some never returned home. It's no wonder there is a distrust among many Indigenous people toward medical professionals. Decades of racial stereotyping and inadequate access to healthcare have contributed to a lack of cultural safety in the health care industry.

I would turn to Indigenous medicine as an alternative to western medicine if I had to choose one over the other. The health industry is so backed up that it's nearly impossible to get in to see a doctor without waiting in long lines. If you have a family doctor nowadays it's like you have won the lottery. Many people in the north do not have access to a family doctor so we have no choice but to start taking matters back into our own hands and be proactive about our health. We need to be our own advocates.

"When you use plant medicine, says knowledge carrier Maurice Zoe, "you can cure yourself." One of the lessons his Elders used to teach him was to drink cooled boiled water. Drink boiled water every day just like you would drink coffee or tea. "Ask the spirit to flush out toxins in your body so that you don't have to depend on modern medicine anymore, because you can flush out the toxins on your own." Water will cleanse you from all sickness and even cancer, he assures.

"You can cure yourself."

Cancer is something that is new for Indigenous people, and Maurice believes that cancer is curable, that the cure has been found, but cancer has become such a big business that the cure won't come free.

One of my children received a cancer diagnosis at a very young age. It was a scary time. I wanted to try and help my child by using Indigenous plant medicine but I was threatened with having my child taken away by social services when I told the doctors that I would prefer to try medicine harvested from the land before resorting to chemotherapy.

Doctors are not trained to suggest alternative medicine, in fact they are taught to discourage patients from using medicine from the land as it may interfere with what patients are being prescribed but that doesn't stop people like Maurice from resorting to plant medicine.

If you have sore bones or muscles the best medicine is spruce sap from the spruce tree, says Maurice. Spruce sap plays a very important role in Dene culture. If someone is sick, boil spruce sap and have them drink it slowly he instructs. It's very bitter but it will help, he says. My grandma always used to tell me to chew on spruce sap but I never did because I didn't like the taste of it. Now I chew it all the time. It tastes like a very flavorful long-lasting gum.

Randy Baillargeon has a story of how spruce sap came to exist as medicine for the Dene. Long ago there were big animals, he says. "Wolverines were the size of a truck." When he was young, his Elders told him that the Dene people were afraid of the giant wolverines. Traps and deadfalls were set but they just couldn't catch the giant wolverines. Then one day Yamǫǫ̀zha came and asked what happened. The people told Yamǫǫ̀zha that the giant wolverines were trying to kill them and eat them so Yamǫǫ̀zha said, 'I'll help you.' Yamǫǫ̀zha went to the wolverine's house and killed the mom and the dad wolverines but the two young wolverines got away. They ran up a big spruce tree and when Yamǫǫ̀zha caught up to them he said, 'Come down from that tree.' The wolverines said, 'No, don't kill us please.' They were scared. 'We want to survive." They cried. 'What do you want from us?' they asked. Yamǫǫ̀zha answered, 'Our people are getting sick…' the wolverine's said, 'We'll help you.' Then the wolverine cubs cried and cried and their tears went inside of the spruce trees. Yamǫ̀ǫzha saw that their tears turned into sap. The wolverine cubs cried, 'Use our tears, they will help you with your sickness.' That's what the Elders say to this day when they see this spruce sap: "This came from the wolverine tears," explains Randy.

When the spruce tree gets old it cracks and the sap binds it. It fills in the tree like an oil says Maurice. If trees don't have sap, they'll die. If you have an ailing body, take sap from the spruce tree, don't mix it with any chemicals, put it in your mouth, it's very sticky, but swallow it anyway he says. "After a while that medicine will lubricate the joints, just like how it prevents the tree from cracking."

I have since replaced my morning coffee with tea and a drop of spruce sap with a pinch of juniper berries for extra flavor. This is not for everyone though. Juniper and other plant medicines can carry risks. Some risks of juniper have been associated with kidney damage and may interfere with certain medications so it's best to consult with a health care provider before changing diets.

Wanda Pascal is the person who gave me the idea to mix spruce gum with juniper. I met Wanda at the women's sewing circle in Iñuuvik. Wanda is a Dene woman from Teet'lit Zheh. She was brought up mostly on the land and sees how it is changing. Wanda was the Chief of her community for two terms, and while in leadership she traveled the world to bring awareness about climate change is impacting her territory. She spoke at the UN about how climate change is affecting northerners in the Gwich'in Region and sharing what her grandparents told her what is going to happen in the future.

When she was ten years old, she was told that in the end the snow on the mountains would melt, there would be no water in her territory, and the ground would cave in—and that is starting to happen now.

Wanda uses a lot of medicine from the land. She uses spruce sap and mixes it with different types of berries. She especially uses a lot of juniper, which she gets from dry wooded areas. She tells me that she helped her uncle, who couldn't walk very well, by boiling the juniper and soaking his feet in the boiled water for twenty minutes a day. Soon enough he was walking better after that she says. Wanda made a concoction of plant medicine for her friend who had a cough and as soon as she drank it the cough stopped. Wanda even switched her own heart medication and replaced it with plant medicine, since then she has seen noticeable improvement in her own health.

When Wanda's granddaughter was diagnosed with a brain tumor, she had to have brain surgery and was sent down south for radiation. Wanda accompanied her, and brought a mix of spruce sap and young juniper with her. Wanda boiled the medicines together for a half hour, then cooled it off and strained it. The spruce sap doesn't taste as strong when mixed with the berries she says and it's true, the berries cut through the bitterness. "About a week after I started giving it to her every day, mucus started draining from her nose." The medicine will suck the infection right out of your body, Wanda says.

When they saw the neurosurgeon he was surprised to see Wanda's granddaughter doing better than expected after taking Wanda's homemade mixture. She was supposed to be bedridden.

"We treasure our medicine."

Wanda's granddaughter is doing much better now but she still drinks spruce and juniper medicine daily as a preventative therapy. People sometimes ask Wanda for medicine, they even offer to pay her but Wanda says that she won't ever sell the medicine that she makes. She gives it freely to those who ask for it. "It's given to me free off the land, I'm not selling it." It's really helping a lot of people. "We treasure our medicine," Wanda says.

It is good that plant medicine can still be counted on, because with climate change westernized medicine will one day be in short supply. Dene scholar Dr. Nicole Redvers is an expert in holistic medicine and fears that the general supply chain of Western medicine will be challenged in the future. Dr. Redvers—who is an associate professor in the Schulich School of Medicine and Dentistry at Western University and is a Western Research Chair and Director of Indigenous Planetary Health—says that despite plant medicine not being as accessible in some regions, plant medicine knowledge may be very useful in the future in terms of increasing levels of antibiotic resistance. As Western medicine becomes more inaccessible, "we will have more people turning toward traditional Indigenous knowledge systems." As an expert in holistic medicine, Dr. Redvers holds Dene teachings sacred and says that bodily health is not separate from the health of Mother Earth. This is the premise behind a fairly new field that's been called "planetary health."

Nicole left the north to attend university, then returned and opened a clinical family naturopathic practice which she ran for over a decade before becoming fully dedicated to planetary health research. Since then she has made it her mission to learn about Indigenous medicine around the world. She now has a master's degree in public health and a PhD from the University of Oxford. Dr. Redvers also co-founded the Arctic Indigenous Wellness Foundation, which was located in a wooded part of Yellowknife, and had the mandate of reviving Indigenous-based healing services and practices in the north.

Nicole always intended to go down the medical route. She was first drawn to working in medicine to help Elders, but after seeing that many people in her community did not seek out healthcare when needed, she realized that a lot of issues were not going to be solved in the clinical exam room due to the structural complexities of the healthcare system. During our conversation I refer to this phenomenon of not seeing a doctor when sick as a deep distrust that Indigenous people have with the healthcare system, but Dr. Redvers avoids using the word *distrust* because it puts the onus on patients and individuals instead of on the system. It's "a system problem, not a people problem."

The way Dr. Redvers understands Indigenous law is that it is the interconnectedness of all things based on value systems and the consequences of not treating Mother Earth the way she needs to be treated. Most of Dr. Redvers's work has centered around planetary health related to climate change, biodiversity and environmental contaminants. In the north "we are warming at three to four times the global rate." The ice melt in the Arctic will affect other areas of the world, but the north is the primary tipping point. Some of the ice that we have held sacred contains mercury and many other heavy metals. When that ice and the permafrost melt, those elements will be released in the north first and foremost, says Redvers.

This rings true to what my great uncle Alfred Baillargeon spoke about in regard to fog and mist carrying sickness. Dr. Redvers says there are "viruses and bacteria that have always been frozen solid in the ice and when that starts to melt it releases the potential of infectious disease. There are unappreciated potentials for many things that can start to creep up when that ice starts to melt." She too has heard from the Elders who already knew this would happen, and spoke of a time when the ground would start to sink into itself because of the permafrost melt—and that process has begun. It is absolutely clear that the amplification of global warming is one hundred percent due to people says Redvers. Those who tend to push back on climate change talk about how there have always been fluctuating levels of carbon in the atmosphere, but it's getting to the point that the last time we had these levels, no humans survived. "It's inhospitable to human life," says Dr. Redvers. "There's no arbitrary discussions around that."

As Indigenous people, we are adaptable and resilient, but "this will likely be the first time we will be tested as a human species from an inhospitable standpoint due to the conditions on earth" if something is not done about global warming. "We're not on a good track right now and we definitely need to be making some adjustments as a society." Dr. Redvers has been trying to inject Indigenous knowledge into national and international policy spaces through her research, to ensure that it is not just there as a tokenization but rather embedded into policy. She has seen some success but there is still a long way to go. Trying to incorporate Indigenous spirituality and ceremony into policy is taboo for bureaucrats. Indigenous plant medicines are categorized as holistic healing or naturopathy by the medical field because there is no category for Indigenous plant medicine and maybe for that reason our knowledge of plant medicines shouldn't be written into policy at all because the lack of respect and open mindedness among health professionals could diminish our sacred ceremonies if they are captured in written format.

"With our medicine systems there is appropriate protocol. There's the usual cold and flu remedies that a lot of families know but there's a different level of medicine where you have to go through certain ceremony." It's not just about the physical act of making and consuming medicine, but also having insight into and a deeper understanding of the sacredness of plant medicine—and recognizing it as a way of life and valuable belief system says Dr. Redvers. A component of the ceremony that Redvers speaks of around plant medicine is about the belief, gratitude and trust that it will work.

These belief systems extend not just to plant medicine but to healing places.

In the early 1990s the Northwest Territories Power Corporation identified a power source at Ts'ąkuı Theda meaning "lady of the falls," but the community of Łutsël K'é was adamantly opposed to the building of a dam because of the spiritual significance in the area—as it is said to be where the Elderly woman Ts'ąkuı Theda will stay for all time. The story goes that the Lady of the Falls was denied beaver blood in the days of giant beavers, and because of that she sits behind the waterfall and waits for people to bring her offerings. If you visit her and offer her something, it is said that she will heal your ailments in return.

In my conversation with Dr. Redvers, I bring up the cleansing pools located across the north that were once used for healing purposes, not many people know about them I say but Dr. Redvers has and says that colonization and residential schools have impacted our understanding of where they are located and how to heal. There is one in Enodah that I have seen with my own eyes. An Elder showed it to me and said that it was for bathing. Possibly as a cleansing ritual that one must go through before visiting the Lady of the Falls.

Elder George Mandeville has come across these sacred healing pools too. He found one near Behchokǫ̀. Having traversed the north far and wide staking claims, when he saw what I believe to be a sacred pool he said it looked like a perfectly round drill-hole, but when he measured it he knew it was not a perfect sphere. "It's like an egg. It's oblong." And it was big enough for a grown person to sit in he says. I remember the one in Enodah was also big enough for a grown person to fit. He tells me about another place where there are perfectly round basins. "They look like bathtubs." There's four or five of them, a series of them next to one another in solid ground. It looks like "someone took a big ice cream scoop and dug them out of the granite... It doesn't look natural."

There are many of these sacred pools carved out in rock across the north that I have been told that we would bathe in them before setting out on a spiritual journey. I do believe that these possible cleansing pools are a clue for us to follow to further get to know our supernatural medicine teachings—and that we, the Dene, need to learn once again how to access them because healing is needed now more than ever.

Dr. Redvers tells me that it's been challenging to find core funding for the Arctic Indigenous Wellness Foundation. But it shouldn't have to be when there's such a need to focus on healing given the amount of health issues and inequities—past, present and future—in the north.

Dr. Redvers says that one of the biggest issues the north is dealing with now is the effects of the oil sands. "We've been pushing for years to have, not only baseline assessments but, ongoing testing for contamination in people downstream from the oil sands." Yet the government and oil companies have refused to fund the research needed to compile that data. The other problem, explains Dr. Redvers, is that when a northerner's health is diagnosed their medical records are aggregated for statistical purposes by region and not by community. This means that health-related data reported by the NWT does not provide localized community data—which presents a problem for researchers who are working to understand how and why certain diagnoses are occurring in one community and not others, to possibly make a case for it being from environmental pollution. Industry cannot be held to account when researchers are not able to build up enough evidence to prove their factual scientific theories.

The argument the government gives for not reporting health indicators by community is due to privacy reasons—because the communities are so small, one might be able to identify people by their personal health information. Dr. Redvers says this reasoning has some merit, however there are deeper issues. "From an Indigenous data sovereignty perspective, the Indigenous communities should be able to own their own data... Communities are basically working in a blind vacuum from a health data standpoint, with the government completely in control over all data-capturing." If you have a contaminant in one region and you want to do an analysis particular to that region or a comparison, you can't. This is the number one barrier when it comes to measuring human health in the north. "Without data it's hard to be able to do any assessments—you can't make an argument. So essentially it voids the community from being able to apply for preventive funding or even research opportunities for epidemiology studies." This means that northern communities can never hold industry accountable for impacts to their health unless the government changes the way data is captured. Even if they wanted to it costs a lot of money to fight against industry, and many Indigenous people in Canada who are being exposed to harmful industrial chemicals and toxic waste don't have the means necessary to hold the industries that are producing these toxic chemicals accountable.

What's really backwards, says Dr. Redvers, is that a researcher from down south can often access community data but the community can't even access their own data. It's frustrating. When the government holds the data, they do what they want with it. They can apply for specific grants and use the community's inequities.

Today, people have a lot of bad habits says Maurice. "The bad spirit takes over them."

The Dene who lived long ago did not consume things that were not healthy. "Back then their bodies were balanced." They didn't need a doctor or a hospital. The healthcare industry only offers a bandage solution, says Maurice. "Our body is balanced by the spirit." Says Maurice.

Dene back then were very healthy and did not die young unless there was a mishap, says Maurice, like if someone drowned by falling through the ice.

There was also no such thing as a eulogy back in the day. People weren't looking for accolades or recognition. "They just wanted to be buried in a peaceful place where their body could go back to nature." We would pray for those who passed and ask them to pray for us too, Maurice tells me. "There's nothing to be afraid of if you live a righteous life," says Maurice, and living a righteous life includes respecting your body and your environment. "If your body is in balance with nature, you will not be afraid of death, not one bit."

I believe that I may have come close to death a few years ago, when I took it upon myself to go out and gather plant medicine. I drove out to the highway, parked on the side of the road and hiked into the trees. I picked flowers and plants I thought I recognized. I gathered Labrador, yarrow, rosehip, fireweed, dandelion and other plants. I peeled the dried sap from the spruce tree. I even collected the small ripe green buds off the alder tree in spring. Back in my kitchen, I mixed everything up with my bare hands, placing bunches of dried leaves and flower petals into small pieces of mesh cloth, which I cut and tied with a ribbon to make a potpourri of sorts. To my dread, I soon realized that inhaling the potent scents had made me feel overwhelmed with dizziness, to the point that I couldn't stand and I became bedridden for the next few days.

> "If your body is in balance with nature, you will not be afraid of death, not one bit."

This was a powerful lesson for me. If I had followed proper protocol, I would have taken the time to ask someone like Paul Mackenzie to teach me how to identify different plants and medicines. I should have also said a prayer and given thanks when collecting the poisonous bouquet. I didn't know what combination of plants paired well with other plants or which plants were toxic in large quantities. For instance, yarrow looks a lot like another plant that is poisonous. I had no idea what types of plant medicines I was working with, and the residuals must have seeped into my skin and my airways—which, I believe, ultimately poisoned me. Mother Earth has such powerful medicines that can be both a toxin and a remedy depending on how they are used.

I shouldn't have gone out into the bush alone. It's always been that Dene women would go out together and pick berries with their children on their back, and they did this so they weren't on their own in case a large animal came along.

When I was working at Dechinta a black bear wandered too close to the camp and had to be killed. The bear was skinned and the meat was prepared to give away to those who consume bear. When preparing the meat, we were careful not to let any parts of the animal touch the ground, as that would be a dishonor to the spirit of the bear. Elder Celine Marlowe was at Dechinta at that time helping the students to tan hides, and she knew how to render the bear fat. She used a cast iron pot, and with her bare hands over an open fire inside of a teepee—while a moose hide was being smoked above her head—she wrung the bear skin so that the fat dripped into the pot until the very last drop. It is a difficult painstaking process but her hands are tough having been used to wringing out hides. Before the grease could harden she poured the rendered bear fat into jars, to be gifted to those in need to use as a medicine salve—mostly for hair and skin, and especially for children who suffer from eczema caused by extreme environmental conditions.

At dinner with Paul Mackenzie, the medicine healer, he tells us of the healing power of sacred animals like the bear and the musk ox. A long time ago, says Paul, a man was hunting. He made a fire and fell asleep while he was cooking a bear, and the hair of the bear burned. When the man woke up he rubbed the ashes of the bear all over his body, and the next day he went back out in the daylight and went hunting. From then on he was never cold again.

Bear and musk ox bladder is good to boil for tea, Paul explains. When a person is young, they should burn the whole head of the bear or musk ox, and once it turns to ash it should be put on the young person's hair. According to Paul, if you do this "your head will never turn white." I wish I had known that before I got all my grey hairs, I say with a laugh.

When cooking the head of a musk ox, when the liquid from the eyes drip into the fire, after poking a needle into the eyes, rub the liquid on your own eyes and you'll never need to wear glasses, Paul says.

My mom who is sitting across from Paul at the table says she made dry meat from musk ox recently. He nods and provides another teaching specific to her needs: "Eat trout heart if you have trouble with your heart." He says with a matter of fact tone. My mom and I both looked at each other as if to say, how did he know? My mom has been having trouble with her heart for a long time.

Paul says when practicing these teachings, always pray for the animal. Eat the heart while it's still pumping. Eat the stomach of the white fish to help cure sickness. He can go on and on with an endless list of remedies. Paul is a treasured trove of knowledge. He stores many of his medicines at his home in T'èʔehdaà and all of his knowledge in his brain. "Welcome to my drug store," he says, whenever someone comes over to ask him for medicine. Everything you need you can get from the land. If someone has a cut, Paul says to put beaver hide on it as a band-aid because it's waterproof.

When Paul gets hungry, he eats spruce sap and birchbark. It's a natural appetite suppresser, "you can lose weight fast and it keeps your heart healthy." "Doctors are still looking for cures, but the old people know things about earth medicine. When you eat earth medicine for the first time it's going to make you sick because it's cleaning up the garbage inside your system."

Paul is a man who knows everything there is to know about healing medicines in his neck of the woods. He knows which medicines are good for curing certain sicknesses. One of them is white lichen, which sticks like glue to large boulders. It is what the caribou eat. It's good for stomach aches or rashes. Not the yellow lichen though, Paul warns—only the white lichen.

The earth's floor contains such potent medicines, and there are as many toxins as remedies, so it's best not to go out on the land and start gathering unless you absolutely know what to do. This is why it is so important to have a medicine teacher with you. I wish I'd known this before I went out foraging on my own, if only Paul was there with me to take me to where he picks plant medicine out on the land so that I wouldn't have gotten sick from my own ignorance.

When I was lying in my bed sickened from my own foolishness, it crossed my mind that maybe I became ill from the inorganic arsenic trioxide that was once emitted out of the smokestacks at Giant Mine, and blanketed the land, never to disintegrate, not too far from where I had foraged. What if I didn't venture out far enough from the contaminated mine when gathering plants? Much of the forest floor in the surrounding areas adjacent to the mine have been tainted from the arsenic that was released into the atmosphere settling on every leaf and branch on the outskirts of Yellowknife towards T'èʔehdaà where I parked my car along the side of the highway and hiked a ways into the woods to gather the plants. Arsenic trioxide is a substance that can never break down. It can never dissolve. What if I was dying of arsenic poisoning? Those were the questions that were racing through my mind as I lay in bed unable to move, for just one teaspoon of arsenic is enough to take a human life.

— 12 —

Generational Justice

The damage caused by Giant Mine is one of the worst mining catastrophes in the world and poses the risk of becoming Canada's Chernobyl. The mine operated for more than sixty years until it finally shut down not long after a lock out resulted in an intentional underground bombing that killed nine replacement workers. Mine management had been struggling with keeping the mine open due to a drop in gold prices.

The mine remains a gigantic monster lying dormant and that is because 237,000 tons of arsenic trioxide dust and 16 million tons of contaminated tailings were dumped underground. Scientists from all over the world have been contracted to figure out how to contain the arsenic safely, so that it doesn't leach into the lake and eventually find its way into the Arctic Ocean. Until a better solution is found, underground freezing chambers will hold the dangerous waste. With permafrost melting at unprecedented rates and forest fires raging exponentially, keeping the arsenic frozen underground is only a temporary fix and will require constant monitoring in the years to come. Forecasted climate change impacts show that, by the year 2080, at least one of the frozen chambers will require buffering or the ice around the arsenic will melt causing it to leach underground.

The Giant Mine site is approximately 900 hectares in size. The area was once a pristine location for bountiful moose hunting and blueberry picking for the Yellowknives Dene, but now the area is nothing but a toxic wasteland. While the mine polluted the air and water, the Yellowknives Dene unknowingly drank straight from the surrounding lakes and rivers. They were warned in English to avoid drinking untreated water from the lake, but many people did not speak English—let alone read or write it. The Dene in the area hauled water from the back bay where the tailings flowed out from a creek named after the prospector who first found gold in the area after seeing it the window of a Dene woman and asking where there was more.

It is not known how many of our people became sick from arsenic poisoning, but when two Dene children died from arsenic poisoning after drinking melted snow that was collected near the shoreline, the mine paid minimal compensation to the families for their loss, $750 per child. Meanwhile, when a well-to-do white family's cattle died from arsenic exposure they were compensated over $20,000. There is a rumor that the streets of Ndılǫ were paved with the tailings from Giant Mine, and I wouldn't be surprised if that were true—especially since the primary smokestack emitting arsenic was named after our Chief Akaitcho.

The government did not want to fund studies to determine our exposure to arsenic from the mine. So we had to take matters into our own hands. A study done by the Indian

Brotherhood back in the late 1970s showed a high degree of arsenic exposure among the Dene living in proximity to the Giant Mine with high levels of inorganic arsenic in their systems—tested via hair samples—yet mine production continued.

A few years ago, I was helping researchers at the University of Ottawa determine if those same arsenic levels could be found in people living in Ndılǫ, T'èʔehdaà and Yellowknife all these years later, long after the mine shut down, as part of a health effects monitoring program. I went house to house collecting saliva and toenail samples, which were sent back to the lab to determine exposure levels. Arsenic did show up in people's systems but the problem with testing for arsenic is that it only stays in the bloodstream for a few days; making the long-term impacts hard to determine.

I wanted to get up close and personal with the monster whose shadow I had lived in most of my life. I was granted a tour and met with government representatives from Indigenous Services Canada at their office building in downtown Yellowknife. Their offices were just finished being built at the same time that the Giant Mine shaft was demolished, which was home to thousands of ravens. The ravens must have followed the culprits, who destroyed their nesting grounds, back to the office because they flew into town by the thousands and decided to take up residence on the new government office building. It was such an anomaly that it made the front page of the newspaper. The building was so full of squawking ravens they had to get contractors to install spikes on the windowsills and ledges to deter the birds from landing. The ravens had come for their retribution.

The woman guiding the tour of the mine had clear thick-rimmed glasses which looked like safety glasses, a red and black checkered lumberjack jacket, and shiny white shoes. As we drove onto the mine site there was a sign that warned of a bear in the area.

The system of signing into the mine headquarters was very old-fashioned. Whoever was on site for the day had their photo identity card up on a pegboard marked "in" on one side and "out" on the other. I had to sign a waiver which warned that arsenic trioxide dust could potentially get stirred up in the wind which posed a deathly hazard. I was wearing sandals not knowing that I was supposed to be wearing closed toe shoes, but the tour guide let it slide. There were a few times that I got out of the truck to take a look around and was told I had to wear a hard hat and reflective vest when not in the vehicle.

We took the old highway into the mine site. I remember as a child sitting in the back seat of my papa's car passing by the mine on that old road. It always smelled like rotten eggs and I had to hold my breath and plug my nose. A lot of the original buildings are still standing but they are gated off and marked for demolition because they are full of asbestos and arsenic dust.

There were a lot of haul trucks going back and forth around the site, and I noticed a few Indigenous women that were driving them. It was good to see that the locals were being employed. We rounded the corner where the freezing chambers were located. You could only see the top of them but they are a series of siphons sticking out of the ground, each one encased in a block of concrete. Nothing about the site looks natural. There are pits and large areas where there is nothing but vacant land.

The old tailings pond was covered over in dirt, and there were robotic machines spraying a blue powdery substance from time to time to keep any residual arsenic dust from kicking up. Dams have been built to keep the tailings enclosed, and there are enormous rusted cylinders the size of a small apartment building that were once used to store water to treat the tailings. The difference between treated water and contaminated water was clearly evident just by looking at the surface of the water in the ponds that were separated by a man-made dam. The contaminated pond was blood red, from a mixture of iron and arsenic that hardens on the surface creating cracks of dry dirt that make it look like planet Mars.

My tour of Giant Mine only served to reinforce what I already knew, that the site is a legacy and the Yellowknives Dene will have to carry the burden of having to warn all future generations because it will never truly go away.

Our Dene stories tell us that giants once roamed freely across the land until they went inside the rocks in the earth. It is believed that one day those giants will return. I believe that time has come. Mother Earth has been split open for the accumulation of wealth which has released a giant so large it can never be destroyed, it can only be held captive for if it were to escape it would be at the detriment of all life on earth all because of the consequences of greed. It will find a way to escape because the second law of thermodynamics is entropy which means that anything contained will always find a way to escape, an isolated system can only increase over time.

There is a story called "The Woman of Metals" in the *Book of Dene.* Long ago the woman of metals had been taken by a band of men to the far side of the great lake. She was held captive for a long time but eventually she escaped. She traveled far and wide across the land to find her way back home. Along her journey, she came upon veins of metal deposits in the ground, copper, silver and gold. They lay scattered in an area where giants once lived. She shared her discoveries with others, but they did not respect her. They violated her, ravaged her and left her for dead not knowing that she was still alive. Wounded and forgotten, she stayed in that place of rock and metals for so long that she eventually began to sink into the ground. First her feet, then her torso, then her shoulders and neck until she was completely buried. Until she herself turned to metal. Even her heart and lungs became metal. She is still there today. She is there for all time. If she is respected, if her body is honored and if a gift is brought to her, she will give metal in return but only if one asks for permission to receive. At all costs, the metal is not to be taken without her consent. Her body is not to be exploited. The metal is not to be taken without reciprocal trade or there will be consequences, for all the pain she endured was not in vain. She is not only the keeper of metals, she is also the mother of the earth.

Between 1948 and 2004, Giant Mine produced 7.6 million ounces of gold. Adjusted to current rates, the overall revenue sits at $4.35 billion. The deconstruction project is estimated to cost the same. The dangerous job of tearing down the roaster in hazmat suits cost approximately $50 million alone.

The Yellowknives Dene will be compensated for what has occurred on our territory. But no amount of money can ever repair the damage that has been done. We are the ones having to live with the effect of the contaminants in our own backyard even though we aren't the ones who made the mess.

A few days after my tour of Giant Mine, I picked up the local newspaper to see smack dab on the front page that there had been a tailings pond spill which may have compromised the local drinking water. Dene Elders always say not to trust the government at face value because even when they look a person dead in the eye and tell them it's safe, they can't promise that what they are saying is one hundred percent true. That is why the words of government officials are not to be trusted because as much as they think they can contain such a beast underground with their engineered systems, the sleeping giant will wake one day and there is no telling how much damage it will cause once it escapes.

Around the same time that the Giant Mine poured its first brick of gold there was an increase in military presence in the north that resulted in a significant environmental injustice affecting many Dene people in the Sahtú Region. Port Radium, now a ghost town, was a town site established on the eastern shore of Sahtú built for one purpose: to extract uranium that was later used to bomb Japan in World War II.

The Dene who were hired to work in the Eldorado Mine and Echo Bay Mine in Port Radium were not made aware of the dangers that they would be subjected to when digging and transporting toxic uranium in cloth sacks, and as a result they fell victim to the deadly consequences. Neither the company nor the government advised the Dene of proper handling procedures, even though they were fully aware that uranium posed a high health risk. They employed Dene men to do the heavy lifting despite the detrimental long-term effects on their health. As a result of handling the deadly chemicals without proper equipment, many, if not all, of the workers died prematurely from direct exposure to cancer from being exposed directly to radiation. For a long time afterwards Deline was referred to as the village of widows.

Throughout the short lifetime of the mine a small town was built up. Dene people who lived in and around the town of Port Radium were segregated from the settler townspeople. There was even a separate school for the Dene.

The book *Bek'eots'erazha Nnide* states that throughout the lifespan of the Echo Bay and Eldorado mines 13.7 million pounds of uranium was taken out of the ground. Dene men worked hard around the clock and made five dollars a day. At no point were they told what they were mining the uranium for. It's fair to say that, had they known they were helping make an atomic bomb to drop on Japan, they would have never agreed to the work.

When World War II was declared over, the settler townspeople in Port Radium got up and left while their coffee was still hot. *Bek'eots'erazha Nnide* says that Dene prophet Ayah predicted that this would take place. After passing through the area prior to Port Radium being established, Ayah envisioned people with white skin living in houses in that area. He saw boats transporting freight. He used his medicine power to travel with them to where they were taking the freight, and he saw that the ore was being shipped to a refinery in the south, where it was being made into a very powerful weapon with the intention to destroy human lives. He saw a large bird in the sky dropping something and that vision was the bomb that was dropped on tens of thousands of people that had similar features as the Dene he prophesized. The first bomb killed an estimated 70,000 people. The second killed an estimated 35,000. When the dust settled, a delegation of Dene people made the long trip to Japan to apologize, even though it was not their fault.

It is not known just how many Dene people died from direct exposure to the radiation because test results were inconclusive and long-term observations of workers were not followed through. But it is proven that cancer is known to be caused by exposure to radiation and most of the men died of some form of cancer. My friend Jennie McPherson's late grandpa Isadore Modeste is one of those men. He had worked at Port Radium, and his family lived out there for part of the year. "My grandfather was always so full of life. He was always busy doing something, but to see him become so frail … it was really heartbreaking."

Jennie is a direct descendant of ʔehtseo Erəyah, Prophet Louis Ayah, through her grandmother Cecile Modeste, whom he raised. Jennie recalls her experience working in Port Radium in her early twenties. She was attending post-secondary school down south, and came home for the summer to work for her nation. It was one of her first jobs. She conducted interviews and collected stories of the people from her community of Délı̨nę who'd lived and worked in the area when the mine was active.

She also collected baseline data to assess the safety of the area. The site had been abandoned long ago but there were still buildings standing. "There were still desks and coffee mugs sitting on the desks and papers flown all around... it literally looked like they got up and left everything." It was really disappointing to see how much they had impacted the environment, and how they just left without cleaning it up says Jennie. "It was disrespectful considering how much money they probably made, for them just to leave, but that's always the case I find when resource development comes up here—they don't have the same kind of mentality as Indigenous people, so they'll just destroy it."

Jennie tells me that she was angry that the government could get away with what they did. "They could care less," she says, because it doesn't affect them—but it affects us. "I don't ever want to see that happen again."

As we sit and visit inside the cabin, Jennie tells me she did enjoy her job in Port Radium, but there was an element of risk to her health. She had to wear a radioactive monitor at specific sites to be sure to stay under a safe level of radiation exposure, and could only spend so much time in those places. "It's scary to think that there's areas around the lake now that even just being there, not even drinking the water or eating the fish or animals from that area, but just being there and walking on it can make you sick, and that's what happened because of mining." Jennie is not against resource development, but if it's going to be done it should be done in a good way—and Indigenous peoples should be consulted and involved to make sure that the land is being taken care of.

The Giant Mine and Port Radium examples of environmental injustice are not isolated. Similar violations of the people and the land have taken place around the world.

Dene protectionist movements have been ongoing for centuries. It took fifty years of speaking out for the oil industry and the government to finally pull back on the Northern Gateway pipeline that would have slithered like a snake across the Mackenzie Valley. Dene leaders of the day like Stephen Kakfwi and Francois Paulette spoke at an inquiry led by then-Justice Thomas Berger explaining that no amount of money would ever be enough in exchange for the health of the land and water because if the pipeline were to burst it would destroy everything that the Dene hold sacred.

Had the pipeline been given the green light, multinationals would have taken gas deposits out of the Arctic Ocean and shipped the barrels to the south to benefit off of, while northern communities would have gotten nothing except the worry and suffering of the consequences of oil spills. And while that project didn't go ahead, big oil resurfaced again in the late 1990s and early 2000s to try to pull the same stunt. Oil and gas executives were interested in drilling along the floor of the Arctic Ocean at the same time that my friend and Dene lawyer Jennifer Duncan just so happened to be starting law school.

Jennifer is very soft-spoken, but she is fiercely brilliant and not to be mistaken for meek and mild. Jennifer is a member of the Behdzi Ahda' First Nation, which translates in Dene

Kədə́ to "caribou-head-shaped point." She also identifies herself and where she is from as Dehlà Got'ine, which means "end of the earth people" in Dene Kədə́. Jennifer has been a lawyer for twenty-one years now, and was inspired to go into law after having many conversations with her mother about the importance of Dene rights when she was growing up.

Jennifer had just finished her undergraduate degree when she heard that the pipeline had reared its ugly head again. "I was really opposed to the Mackenzie gas project mainly due to the concerns about climate change and how the development of natural gas would contribute to global warming." She knew it would be detrimental to the region and the planet. So she took the reins. She started a non-profit organization to oppose the pipeline just as the environmental assessment of the pipeline was underway.

"We organized the opposition to the pipeline."

The Arctic Indigenous Youth Alliance was co-founded with her friend Elaine Alexie, who is Gwich'in. The alliance brought together young Dene and Inuvialuit people from across the NWT to stop the pipeline. "A lot of the Indigenous leaders were seriously considering saying yes and being courted hardcore by the gas companies." The idea behind starting the non-profit was to not be a lone voice for political leaders to attack. Jennifer laughs when she looks back on it. "We organized the opposition to the pipeline." At that time there were a lot of climate change deniers, and big oil was spreading propaganda that the pipeline was nothing to be concerned about—which she says made the alliance have to work harder to educate and bring awareness to the communities about how that wasn't the case. There were a lot of advocates for the cause among the Elders, who encouraged her and other youth to speak up against the project and eventually the group created enough pressure and resistance to slow down the pipeline by working alongside other groups across the north during the tribunal process. This paired with the fact that approvals weren't given as quickly as expected caused the pipeline to slow to a halt.

It was a delay tactic similar to the Berger Inquiry when the Indian Brotherhood stopped the first pipeline. Justice Thomas Berger was the man sent in to hear the concerns of the Dene. "I actually sat down with [Thomas] Berger himself and asked him... if he had any pointers or tips or anything he could provide in terms of strategy." Says Jennifer. When the goal of stopping the Mackenzie pipeline project was met, the Arctic Indigenous Youth Alliance dissolved.

Jennifer's next big crusade after the pipeline shut down was negotiating in Paris while studying French. She knew it was important that Indigenous voices be included in the Paris climate accord negotiations and found a way in. "Advocacy at that level is just so intense. It's so hard to get into those rooms but it's so necessary, and I think there's a lot of great efforts happening now especially with enhanced participation by Indigenous people." She worked her way into those rooms to make sure Indigenous voices were amplified, with the focal point being the protection of northern lands and water. Even at that level, there were times when she doubted if she was even making much of a difference. Her efforts to strengthen international support and gain momentum to bring the issue of Indigenous rights and respect for the climate to the table were brought forward.

After advocating tirelessly in back-to-back pursuits for environmental justice, first the pipeline then the Paris Agreement, Jennifer felt drained. The work was traumatic even though it had been a success. "I had to take a step back and focus on my own spiritual, physical, mental health, and get grounded again. It can be a very violent process." Your whole identity can be torn down or diminished, she explains. "You are really opening up your mind, your heart, your soul, your whole existence to plead for some attention and protection, so it can just be very difficult to go through."

We have to keep kicking down doors and repeating the same messages over and over for them to finally hear us, which can be exhausting.

"We are talking about the land and our relationship with the land and animals and trying to explain that to non-Indigenous people... it's just dismissed or ignored."

Jennifer went back to her home community and went on the land on the heels of the accord. She was always taught that the land is where you can heal, and she says, she's fortunate that she can go home and go berry picking, check nets and hunt with her family anytime she wants. It is restorative and reenergizing for her spirit.

"After I recovered," she laughs—having to make light of the seriousness of it—she had moments where she saw that her advocacy did make a difference.

The biggest culprits for producing greenhouse gas emissions are the electricity and heat production sectors and the transportation industry. The burning of fossil fuels and the development of hydro dams are major contributors to greenhouse gas emissions but the agricultural industry is in a field of its own.

The north is not the best location for farming for obvious reasons. It is rocky and the weather is inhospitable for certain conditions to raise animals and grow healthy crops. But agriculture is not just about growing fruit and vegetables anymore. Modern agriculture is pumping animals full of hormones, then having to mitigate the deadly biological viruses that come from the poor treatment of animals. Then there is the engineering of genetically modified organisms that use harsh chemicals that contaminate natural crops in the process.

When European farming techniques began to expand in Canada, farming equipment, livestock and seed was supposed to be provided to Indigenous people in exchange for government access to land through the treaties, but that promise was broken like many other governments promises.

Because the Crown did not hold up their end of the bargain, an agricultural benefits settlement agreement has been reached whereby the government now has to pay each and every Indigenous person a dollar amount equal to the estimated loss that occurred because of the failed promise including the loss of land. This agreement has been nicknamed "cows and plows."

Indigenous people had their own methods of cultivation long before the treaties and did not need European farming influence, but when they were denied the ability to leave their reserves to hunt, fish gather and fertilize their crops they had no other choice but to rely on government handouts.

In the late 1800s and throughout the 1900s, government officials, then referred to as "Indian agents," were sent to communities to systematically deny food rations to Indigenous peoples, by not allowing them to leave their homes to go out hunting. This was compounded by a game reserve system that was created in the NWT in the late 1800s, where it was prohibited for Indigenous people to hunt, fish or trap out of season. It is through the game reserve policy that hunters like Michael Sikeya were prosecuted for hunting on their own land to put food on the table.

Full telecommunication did not come online in the north until the 1970s. Before then, channels of communication were limited and some Indigenous people were fined or jailed for hunting and were not able to warn other hunters about what was happening to their Aboriginal rights fast enough or gather together to protest the act. Many hunters ignored the game rules; they had to, or else they would starve. When they were caught they were not able to defend themselves because they were prohibited from seeking legal counsel. It wasn't until the 1950s that Indigenous people were allowed by law to access a lawyer.

Meanwhile settler hunters and trappers were using poison in their bait—chemicals like strychnine and inhumane trapping equipment. The Dene did not agree with the practice of using poison and were in fact afraid of using it.

The game reserve policy proved hard to enforce and was eventually rescinded, but it is just another shameful example of the government's unjust tactics to gain control over Indigenous lands.

Growing up with my grandmother, I don't remember eating very much store-bought meat. We always had a freezer full of caribou or moose meat and fish. Sometimes we would have the odd duck or rabbit. Today my family might be lucky if we have wild meat once a month or a few times a year. It is a rare treat nowadays to have fresh, locally harvested food on the table. This is where communal sharing is crucial.

Many northern communities are relying more and more on expensive food shipped in from down south. This poses food security issues in the face of climate change. when the barges bringing in food to the communities via the Dehcho are not running anymore due to low water and the risk of getting stuck on a sandbar, making food even more unaffordable to residents as a result. Low water causes food to have to be flown in, which raises the price of an already high-priced food market, and because there are only a few stores, often just one store in community—the Northern Store being the monopoly—there is little to no retail competition.

Łutsël K'é not having their regularly relied-upon food shipped in during the Yellowknife fire evacuation is an example of the extreme food security issues that could become more and more prevalent in the north as global warming continues to make an impact. When visiting the community, I was told by the community guide Stephanie Poole that when the city of Yellowknife was evacuated a few years ago, Łutsël K'é was nearly forgotten. When the planes stopped flying, Łutsël K'é was unable to rely on their regular shipment of food.

The odd plane that did deliver food landed and took off immediately, as though it was dropping food rations to victims of the war. The community had to scramble together to unload the package and share it equally. The shelves became empty in the one store in the middle of town when people realized they would be cut off from their regular food shipment. Thankfully food was donated by local volunteers who had harvested the community gardens in Yellowknife, and the Co-Op organized for a package to be put together for each person which was delivered on a private charter.

Many people in far northern remote communities were forced to rely on shipped-in food because they were historically relocated by the military so that their traditional homelands could be used for commercial fishing enterprises. The Inuit were relocated to areas where local food sources were limited and extreme unsheltered weather conditions prevented them from being able to fend for themselves. This has not only caused food security issues, but also a lack of socioeconomic prosperity to this day.

Although Tuktuuyaqtuuq is not one of those communities that was relocated, it may soon have to be because of coastal erosion caused by climate change that is predicted to happen by 2050. Three houses in the community have already been moved further inland to prevent them from being swallowed up by the sea. In my short visit to Tuktuuyaqtuuq I saw just how surrounded the community is from all sides by the Arctic Ocean. The wind is a constant force there, and one big storm surge could wipe out the whole community. In Tuktuuyaqtuuq, most houses have two storm doors as an extra precaution, and most of the buildings are hoisted on stilts to prevent the heat from the homes melting the permafrost and potential flooding.

Historically, relocation tactics in Inuit communities along the coast have resulted in starvation, because of the lack of access to food sources in environments where food is scarce, which has led many people to fear further relocation because of the trauma that was inflicted on families as a result. Inuvialuk archaeologist Letitia Pokiak is conflicted on the topic of relocation in Tuktuuyaqtuuq. Some people think that infrastructure like a sea wall should be put in place to protect the community from coastal erosion, to buy themselves more time. There is a sense of nostalgia, this is where they have settled, she says, but in the past people would have just moved further inland because long ago people were nomadic or semi-nomadic. Times have changed and approaches to climate action have adapted, and the town will ultimately need to be relocated.

There may have been bad years of starvation when our ancestors were nomadic, but we were always able to rely on the animals, says my cousin Tommy. We followed the caribou. We knew when and where the fish were plentiful. If the fish weren't biting, we'd move to another side of the lake, says Tommy. It was that simple. But it's not so simple anymore. Sure there were times of famine in the days before currency and the introduction of westernized diets came into the north but the introduction to new foods also has its challenges. Flour and sugar became household staples when trade routes began in the north between Indigenous peoples and Europeans. A certain type of bread called bannock was born out of the trade era and is now a culturally accepted staple in many Indigenous households. Certain vegetables and fruits that weren't native in the north were also brought in, and eventually Indigenous northerners began including new foods into their diets even though they were not natural to their environment. Fast forward to today and fast-food culture is causing growing health problems such as diabetes, obesity, dental problems and heart attacks among Indigenous populations.

There have been conclusive studies done in many areas of the world that show that nutrients in food are decreasing as the heat rises. In her work on planetary health Dr. Redvers urges us to see food sovereignty as a global warming issue that needs to be included in climate change discourse.

So, what exactly is food sovereignty? Well, it can look like grassroot community gardens for one. There are local community gardens set up in many communities across the north that provide produce for free to residents. I visited the community garden in Iñuuvik and was surprised at how abundant it was. A diet of meat, fish and berries is what the Dene have relied on for centuries; but a diet of assorted fruit and vegetables could fill the gaps when locally harvested food might be in decline or unavailable. But nothing can replace locally harvested food that is natural to the environment.

The GNWT has funding set aside for regional harvesting, a training and mentorship program. They also have a pilot program for trapper mentorship to assist families in going out on the land to put food on the table, but the amount given is not enough. The funding should be radically increased so that these programs have the furthest possible reach. Hunters should be paid to help their community. Many Indigenous residents in housing are unable to afford groceries after paying the high cost of rent and utility bills and rely on hunters in the community to provide. Harvesting programs need to be supplemented seasonally across the north to support the Aboriginal right to hunt, fish, trap and gather. Food security is an Aboriginal right in the face of a changing climate. Funding for Elders and seasoned hunters and gatherers should also be provided so they can train the next generation of hunters and gatherers. Understanding the changing environment, and having the skills to be able to culturally harvest local foods and process that food, is necessary for northern food security. Without proper training inexperienced hunters can become susceptible to accidents when traveling out on the land. New hunters need to know how far they have to travel to harvest their food in case plants and animals are exposed to pollutants. The extent to which Indigenous knowledge is transferred through the generations plays an important role in determining the health and wellness of individuals and communities.

Revitalizing Indigenous food security through local harvesting can help communities prosper. Innovative and sustainable social enterprises in food manufacturing and distribution should be set up in the north to access locally harvested food. However, the GNWT has so far made minimal effort to work with community governments to support the opportunity to bolster food security, and instead a lot of money is going into the high cost of importing international foods and supporting international trade even when they are fully aware that perishable food items being sent to the more remote communities are nearly rotten by the time it arrives. The domino effect is that this encourages the use of more and more pesticides and preservatives. The shipment of food to remote locations is a major contributor to greenhouse gas emissions, but trying to stop international trade would cause an economic crisis—and broken trade agreements could lead to conflict on both a small and large scale. The agricultural industry is locked into international trade agreements. If we stopped mass production of food altogether, conflict would be exacerbated, which could lead to further wars and devastation globally and the high cost of gas at the pumps would be the least of our worries.

I walked into a health food store once and saw frozen kangaroo meat for sale. This is not necessity, it is a trend. With the click of a button, food is being shipped across the ocean in a container and landing on our doorstep.

I am guilty of trying to transport food myself. I crave caribou meat when I don't have it and so a few years ago, when I was fortunate enough to be given some caribou meat before heading back to law school after summer break, I brought the caribou meat with me. It's about a three-day drive to get back down south from Yellowknife, so I packed the meat as best I could in a cooler filled with ice. When I got into the mountains two thirds of the way it started to melt as I expected it would. I got to my hotel and unpacked the meat and put it in the bathtub. I refilled the cooler with ice from the ice machine, but the ice machine broke and the meat nearly spoiled but I made it home the next day just in time to put it in the freezer.

When my family came to visit me one year, they also tried to bring harvested meat in their luggage and checked it in before getting on the plane but they ended up getting stuck in a "polar vortex," which shut down a lot of the airlines. They were stuck in a hotel over the Christmas holidays while their luggage sat in the lost pile among hundreds of other suitcases. When we finally retrieved the luggage we had to throw it out, meat, clothes and all, because the meat had spoiled. We go to great lengths to have locally harvested food, but sometimes it doesn't always work out in our favor.

I once heard an Elder say that when two animals from opposite ends of the earth meet on the same plate it will be a time of great change. We see this happening now with international trade. Why purchase blueberries from a country across the ocean when they grow naturally on the land? I understand that international trade helps struggling countries in a fluctuating market, but now that we are in a full-blown climate crisis we should try to only eat local and fill our baskets only with what is in season, even if that does mean that jobs will be on the line and it goes back to the reason for eradicating money altogether.

Robin Wall-Kimmerer, in her book *Braiding Sweetgrass,* calls eating locally an "honorable harvest" and includes a set of guidelines that include taking only what you need, taking only what is given, never wasting, always sharing, giving thanks and reciprocating.

Former Chief Wanda Pascal lives by the code of the honorable harvest by relying primarily on locally harvested food. She tells me that she wouldn't have a problem having to live completely off the land because she has access to it and she is thankful for that. Wanda's grandson would rather go hungry than eat processed food because he lives off a diet that mostly consists of locally harvested food. She was told by her Elders to only eat a land-based diet as much as she can, because in the future people would have a hard time affording their next meal and that time has come. Grocery bills have become sky-high especially in the north.

The northern food supply chain has been put to shame on an international scale. Photos of outrageous prices in Arctic communities have circulated on social media and in the news—yet nothing has changed, even after scrutiny from the auditor general of Canada, who oversees independent fact-based research on government programs.

When in Iñuuvik, I wanted to compare the price of steak to store-bought steak down south, but there was no comparison because the cost of a steak in Iñuuvik was so outrageous that nothing could match it. Two steaks, priced on sale, rang in at the till at fifty dollars, not including tax. A pack of a dozen bottled waters can be as much as fifty dollars in Iñuuvik, and wouldn't be a surprise if they were filled with tap water. Toilet paper is so expensive that it wouldn't be that drastic to consider going back to using moss from the ground in its place.

The northern food supply chain was once owned by the Hudson Bay Company until the Northwest Company bought it out. The companies now hired to ship food to the north are being paid out a portion of government subsidies, when those subsidies should be going directly to the people living in communities who can't afford the high cost of groceries. The northern food subsidy program, Nutrition North, was originally intended to help mitigate the high cost of food in the north, however Nutrition North Canada has been under the microscope for some time now. The lack of community confidence in the program stems from the suspicions that northern retailers have worked out a back door deal with the airlines and barges that are shipping the food, saving the companies money while the cost of food for residents remain high and subsidies remain low.

Foods that are subsidized under the program are categorized as perishable and non-perishable. Whether one receives a low, medium or high subsidy depends on a person's income. To make matters worse, food subsidies are factored in as income and taken out of social assistance payouts.

Subsidized food vouchers are simply not enough. Food banks and churches, run by volunteers, are helping where they can to alleviate the cost of food by providing food and vouchers to people in communities who are in need—but there is no promise of regularly receiving this help. There is no continuity because the shelves of food banks are often empty. Try as they must, these programs are not reaching the masses. Many Indigenous people are not inclined to ask for anything from the churches because of the dark history of the residential schools and the distrust that still exists. Not to mention that the vouchers are only perpetuating the purchase of store-bought foods and not putting resources into making locally harvested food more accessible.

Locally harvested food was never bought and sold but some people sell it, yet many people are still of the belief that it should not become a commodity because the animals might not return. Wanda Pascal does not sell wild meat, instead she gives it freely to the community when she can. She believes that if she shares, she will get it back in return. This goes back to the Dene law of sharing what you have, yet sharing is becoming difficult for some who are worried about their next meal.

Many hunters now only hunt for their immediate families, whereas in the past hunters would go out together on the land and share whatever game they caught with the community upon their return. We still have community freezers, but sometimes there is not enough food to go around. Elders and children are given meat first, which is good. And community freezers are emptied before the fall hunting season begins, because it is believed that if you already have meat in the freezer the animals will not present themselves to the hunters because the spirit of the animal would know that they were taking more than they needed. Sometimes, though, by the time the meat is ready to distribute to the community, it is freezer-burnt from sitting in the freezer for too long. It is a shame to have meat go to waste. This goes back to the Dene Law of only taking what you need.

In the past the Dene did not hoard for years to come like we do now with our pantries full of canned goods. We only ate fresh and locally season to season. Meat was only stored for so long because there were no freezers, just the cold hard ground or lake. Either that or it was cured and dried to keep longer. If someone still had meat stored by the time the next hunting season came along then they would give the meat away before going out to hunt again.

Implementing locally harvested school food programs is one way to address food security concerns in the face of climate change. This would help mitigate the need for imported foods while also preventing long-term health-related impacts from processed foods like diabetes and tooth decay. Food security associations and existing food-growers should be open to partnerships with Indigenous nations to offer different programs such as school lunches that encourage eating locally harvested food and teaching children how important it is to eat a healthy diet while also caring for the environment.

We must do our best to ensure that our generations are healthy and have a healthy planet to live on. It is often said that we should listen to what the youth have to say, but in a world of fast changing politics, young people's voices are often left out. Yet, when we put our trust in youth to get the job done, amazing things can happen. Just look at Dana Tizya-Tramm who is the former Chief of the small fly-in community of Old Crow, Yukon, who led the way in establishing a solar farm in his community. As a young Chief, Dana sat at negotiation tables and ultimately got the Yukon government to agree to creating the independent power producer policy. He was not jaded or held back with the pessimism that adults often carry, perhaps youth bring a sense of naivety that is necessary to do the impossible.

I happened to be visit with my friend Cassandra Blondin Burt the day after she got back from a community tour across Dënéndeh as a musician and storyteller. Around the same time, she was hired to facilitate youth climate change surveys in the schools in one of the regions. I visited her in her off-grid home at the bottom of the hill by the shoreline in Ndılǫ. Her house is equipped with everything she needs to live off grid. Cassandra is sitting in the dark with the woodstove blazing. An assortment of animal bones are strewn across her dining table, and in her kitchen are cooking pots full of plant medicine that she is packaging to give away to people, as well as salve and bath salts to sell. Cassandra works with plant medicine. She is also a spoken word artist and author, among many other talents.

Cassandra had some interesting observations to share with me from her school visits. "What I realized very quickly was that young people didn't just need more opportunity to speak. They needed to be given the space to effectively share their insights on climate change policy." Some of the youth were very quiet and what became obvious to Cassandra after some deep reflection is that silence can sometimes be an expression of protest in itself. Protest doesn't always need to be loud. "Here me out," she says excitedly. "I take their silence as a rebellion." To break that silence, youth need to be given opportunities to learn and practice effective communication skills and have access to safe spaces where their insights are honored and actioned upon so that there is no room for fear.

As it stands now, many youth are experiencing extreme climate anxiety.

"If I try to translate climate change into my worldview it doesn't compute. It literally doesn't make sense." We are not facing climate doom and disaster; we're not facing an end time, says Cassandra. "What we are facing is a dark night of our collective soul." Centering this in Dene methodologies would be "like a young person who has been put on the land alone as a coming-of-age ceremony." Now if that young person is out there is living in a narrative of fear, says Cassandra, then they're going to feel alone and afraid that they are going to die, thinking "I'm never going to survive this" or "I'll never make it out alive." It's all in the way we see the world, she explains.

How would we want our young people to be when alone out on the land? she asks me and before I can answer she says, "We would want them out there knowing that the dark is nothing to fear, that they are prepared, that they have everything they need and in fact, even though it is dark... the sun will rise." It's about wisdom and best practices, not having the best tools, technologies and the most expensive things. "It's going into the night knowing the dawn will come again. It's going into the night knowing that your mind is what creates the environment you exist in." We want our young people to go into that dark night, pushing through the feeling of fear, acknowledging the reality of the danger of the situation and the gravity but having the ability to stay calm and grounded within that—because if they stay calm and grounded then they can "set up camp, they can make a fire, they might even be able to hunt for themselves," Cassandra says. If they have it in mind that they are going to make it through until the morning, the sun will rise and they will not only be okay, they will have come through that dark night with strength, integrity and honor, and they can come out the other end with pride. "It's not fear that gets you out alive." It's wisdom says Cassandra.

During her tour of the schools talking to the youth, there was resistance that came with having to answer questions around climate change because of the way it was framed, so Cassandra changed the question. "They didn't have any response when I asked, 'How do you feel about climate change? Do you want to become a climate change activist?' Most of them sat there silent and kind of glared at me," she laughs, "but when I asked 'How many of you really love being on the land with your grandparents?' they all put their hand up. When I asked 'How many of you want to be able to do the same thing with your children and your grandchildren, and how many of you are willing to work very hard to ensure that that happens?' all of them put their hands up. That's a perspective that makes sense to their spirit." And this is what generational justice is.

The current Western narratives about climate change are quite false says Cassandra. "Not only are they false but they create division, they create the illusion of ending, of apocalypse narratives." This then brings about in our physiology lower basal responses because everybody's in fight-or-flight mode. Whereas from an Indigenous worldview it is much more about the earth cleansing herself and that feels much safer than the end times explains Cassandra.

Us regaining our sense of wisdom by operating out of an enlightened or elevated frame of mind means moving out of the fear narrative in order to embrace a more holistic view of just how intelligent Mother Earth is, says Cassandra. This is not an insight that she arrived at overnight. Cassandra has gathered this knowledge from everything that her great-grandpa George Blondin's books teach. This has been understood for thousands of years she says. "Whether it's making my plant medicines, or doing fire research, or sharing the stories, I believe that if I'm honoring these stories, if I'm living in this truth, my physiology changes, therefore my neurobiology changes, therefore my experiences in my daily life change." Cassandra, being a young person herself, says that when she is no longer in that space where she has to fight against some unseen rising tsunami that is the climate crisis then she is able to "sit in concert with the universe"

Cassandra's wisdom is years ahead of her time. When I tell her she'd make an excellent professor with her lectures, she responds by saying she feels like a ragamuffin running around in the stars.

There's a lot of fear right now and that is why her storytelling is intentionally soft and loving, because for her it feels like the whole world needs a mom. "Everybody needs to be told that it's going to be okay and that we can trust our mother. We can trust her to transform this world for us, but we can't be fighting with her about putting on our shoes."

— 13 —

Time Immemorial

On three separate occasions I have had the opportunity to visit an Elderly couple named James and Elsie Shay, who are well-known healers from Dinétah, the home of the Diné. The first time I met them, they were visiting the community of T'èʔehdaà and offering guidance to community members. I had just started writing my first book but hadn't told too many people about it. I sat across the fire from them and the first thing they told me was that I would write a book one day.

The second time I met with them was when I heard they were visiting with community members at the Wıìlıìdeh site during the summer and I went to see them. I brought my daughter with me, and she visited with them inside the teepee first. They asked her what she was seeking guidance for. She told them she was thinking of moving away from home to go to university to study anthropology, and they told her she would be protected. When it came to my turn, I went inside the teepee and sat on the chair next to James who made small talk with me as we waited for Elsie to come back from refilling her coffee cup just outside the tent at the cookhouse.

"Where are you from?" he asked.

"From here," I said and told him I was Yellowknives Dene.

He looked at me sideways and said, "You look like a white woman." I laughed. He didn't remember me from the time I had visited him a few years ago. I joked and said I'm like a shapeshifter, I'm always changing my hair color. I told him that my father is a redhead from the east coast but I was not raised by him and grew up in my grandma's Dene culture raised by my grandma, and he went easy on me after that.

When Elsie returned to the tent and new coals were added to the fire they both asked me what I was there seeking guidance for. "Well you told me I was going to write a book one day and I did and now I'm hoping to get some advice on a new book I'm writing about climate change from a Dene perspective. I'm halfway done writing it and have more questions than answers." I added.

Elsie opened a bag of tobacco and tossed some onto the hot coals of the fire while James stirred them around with a stick. "So you want to know if the book will be successful?" Elsie asked inquisitively. Not quite, I said, "I just want to make sure that it's done right. That I'm following proper protocol." James then picked up an eagle feather from the ground by his feet and put it in my hand. "Hold it like a pen," he said. I did as I was told. "From now on when you write you will be guided by the eagle."

What I know of the eagle is that it is a visionary, and I've always felt that the eagle leads my way. Then James blew into a creamy white hollowed-out bone that looked like a straw. It made a loud whistling sound and a strong gust of wind suddenly blew into the teepee. The small blanket covering the entrance flew open, and ashes from the fire floated all around me—covering my hair and clothes in white powder. "The ashes will protect you," said James.

It made sense. My grandma always said ashes have healing properties, and hearing Elder Paul Mackenzie and Gilbert Cazon talk about the power of ashes, and now James sealed it. I thanked them both and before I left the teepee James nodded and said, "Your work will be respected."

The third time I met with the healers was on their home turf in Dinétah.

It was John B. Zoe, Maurice Zoe's brother, who encouraged me to take the trip. I called John B. one day to ask him about the hole in the rock near Whatì that Chief Fred told me about and how it was a portal that the Tłįchǫ walked through. I knew he would know more about it being that he is a Tłįchǫ knowledge carrier. He is from Behchokǫ̀ and was the Chief land claim negotiator for the signing of Treaty 11 self-government—among a long list of other accomplishments he has earned over his lifetime, including receiving an honorary Doctor of Law.

John B. answered my phone call while he was driving in his truck on the highway and I could hear the dinging of the seatbelt warning in the background, or maybe it was his turning signal I couldn't be sure. I talk to him about my thoughts on how and why the Dene separated from the Diné long ago and that I was informed by others that it was indeed for our own survival. To that he says that it has been confirmed that Dene history connects to the Diné through undisputable evidence. "If we count from one to ten, we can understand each other. Our languages are about ninety-five percent the same."

He says that our stories of the sun, the moon and the water are similar to those of our distant relatives as well.

I tell John B. that I've been thinking about going to a place called Window Rock on the Navajo Nation reservation, and he's already way ahead of me—having visited Tséghádhoodzání before I was born.

Is Window Rock connected to the rock near Whatì? I ask. First, he says, in between working, eating, sleeping and spending time looking at screens, if you want to understand more about who we are, it means going out on the land for long periods of time.

Go to Window Rock, speak to the rock and ask the rock for answers he instructs. He tells me he just so happens to be in the middle of writing an article on how rocks can speak so it's good timing that I'm asking him about Window Rock.

Just like in the desert, there are many unique rock formations across the north, but the one near Whatì is the most popular because it is an unusual rock formation he explains.

There is petrified wood inside the hole near Whatì, says John B. There is a lot of room for interpretation of how it came to be. He does not veer away from the scientific evidence, however. As much as he is of a logical mindset, John B. still possesses the power of expression through creativity. In his lectures or in one-on-one conversations about politics, he often draws elaborate sketches of governance systems that look more like a piece of art that should be hung on the wall of an art gallery or museum. One day I hope to own a piece of this unique artistry, signed in the corner with his signature. Nothing short of a masterpiece.

And just like people are free to interpret a beautiful artistic masterpiece differently, John B. says, today we are able to look at the rock through our own lens—whereas for the past 100 years we were prevented from looking at the world from an Indigenous perspective. We had to look at the world as wards of the state, while our lives were being wiped out and converted. Part of reconciliation is revisiting our own histories, relearning our languages and reestablishing a connection to the land, and how we do that is journeying for answers. There is no shortcut to finding the truth says John B. And with that I set out to seek the truth of whether the two similar rock formations in our distant territories are connected.

The second I got off the phone with him I booked a flight for my daughter and me, at a time of heightened turmoil between Canada and the United States—because of the trade war politics raised by the Trump administration.

When my daughter asked me why we were going to Window Rock, I told her it's because I needed to go talk to a rock and she looked at me sideways. "Just trust me," I say. And I repeat what John B. told me, "When in the hands of the land, the only thing that the land requires is that you talk to the animals, talk to the plants, talk to the rocks and then listen."

I let the healers know that we would be coming their way for a visit, in hopes that they would make the hour-long trip from their home in New Mexico to Tségháhoodzání, Window Rock, to meet us—and they gladly agreed to meet us there.

On the long drive across the Arizona desert, the scenic landscape reminded me of home. There are ravens in the Navajo Nation as big as the ravens we have in the north and the sand gently drifting across the road in front of us looked like the snow in winter in Yellowknife how it travels low to the ground in the slightest of breezes. Even though we are on the opposite ends of the spectrum temperature-wise, our lands are both deserts—it's just that ours is a cold desert in the north, and theirs is a hot desert in the south. Both are dry, and the skin cracks around my fingernails the day I get there.

When we got to the town of Window Rock we headed straight to the local McDonald's which was our meet up spot just down the road from the actual Window Rock land formation next to the Navajo Nation band office. We sat together in the busy fast food restaurant sipping tea and catching up. One of the McDonald's workers came over to us and cleaned off our table with a cloth. She smiled and I had to do a double take because she looked just like Elder Celine Marlowe.

I started asking James questions about Window Rock and whether or not we could walk up to the top of it, and he said if we tried to go unaccompanied by a ranger we would probably be shot on the spot, so fair to say that was out of the question. He started in on telling us a story about a set of twins and how they walked on a rainbow down into the valley where Spider Woman lived. It was difficult to understand him at times because he spoke softly out of one corner of his mouth. James is an Elder with hearing aids, a cane, and noticeable cataracts under his glasses. His brown eyes are turning blue. When he was finished telling the story of the twins, he told us that it was only a portion of a much larger story and that we should not be talking about it inside, especially in a McDonald's of all places, so that was our cue to head out to the site of Tségháhoodzání that the town is built around and it was there that James said a prayer in his language as the warm wind blew tumbleweeds around.

When it was time to part ways, they suggested we go check out the Navajo Nation Museum and we did. The museum had a library where I could have spent countless hours researching, but we needed to get back on the road before dark—and besides, my daughter wanted to go check out the Navajo Nation Zoo just down the road. The zoo had all kinds of animals. They even had a Canada goose. The animals were all rescues, most of them amputees, and having them taken care of in a closed environment was their best chance of survival.

In visiting Dinétah and talking to the rock I received answers to some of the questions I had that I wouldn't have gotten had I not gone there in person. Still, as I suspected I would, I was left with more questions than answers, and it might take me a lifetime to figure out if the rock formations in both our territories are somehow connected. What I do know is that some believe that they are portals that were once used to travel long distances in short periods of time at a time when medicine power was at its highest form. A time travel vortex.

This makes me think about the story that Randy Baillargeon shared with me about time travel. Randy, the goose hunter, told me there are two hills near T'èʔehdaà where people used to go. At that time there was no town, no stores, we lived off the land, says Randy. There were three healers who went up on one of the hills. One of them was very strong. He went into a tent and the two other healers stood outside on either side of the tent. The strong man started to shake. "His whole body was shaking. That meant he was ready to travel the world," says Randy.

The strong man's hide-tanned clothing was inside out. "He had birchbark on his face." He wore it like a mask. "That too was inside out." When the two other healers let go of the tent, he vanished. The people asked the two other healers where he was going. They said he was going to check around the world for what is going to happen in the future and so the people waited for him to return. "At that time, there were no clocks. They looked at the sun. 'When the sun goes down he will come back,' they said." When the strong one came back his body was very hot and he said, "People will come." The people asked how many? He looked up to the stars in the sky and pointed. "That is how many," he said.

Dr. Redvers says most people who are not exposed to this type of information may have a difficult time appreciating it. The three healers in the story that Randy tells were in ceremony and the power of ceremony, says Redvers, is being able to go into states that are ancestral. "Many Elders still have those ceremonial processes. The concepts that are mobilized through those processes are part of the visioning that occurs."

The Dene used to time travel in their minds. At one point when medicine power was strong among many, Dene didn't even have to talk, we just used our minds and energy to communicate—but because we have been given phones and other tools to communicate, we have become distracted. We need to break through the screens that act like a veil that has been placed over our eyes.

Both man and animal once had the power to time-travel. In *Yamoria the Lawmaker*, George Blondin says that the wolverine had the power to fold the earth in half like a piece of paper. For each one of our steps, the wolverine could take one hundred. George Blondin's great granddaughter Cassandra Blondin Burt believes that our Dene ancestors did not need to use anything other than their minds to get to where they wanted to go.

"When I'm reading my great-grandfather's books, I'm not just reading books about the stories of our ancestors and the great things that they did, I believe that I am reading a textbook that shares a Dene methodology of living in balance with nature." Yes, we did make tools and we had technology, but what was more important was traveling light—so we needed tools that were not heavy to carry. Cassandra believes that the tools we now have at our disposal are the stories that have been left behind for us—the stories that were shared by our ancestors that were given to us to use as tools. What could be lighter than a story that lives forever in your heart and mind? She asks.

When in a staff meeting once, we went around the room answering the question of what we would do if we had a time machine. Most people said they would go back in time to see the dinosaurs or stop certain events from taking place and warn people. Only a few of us said we would go into the future. Imagine not needing a passport, not having to wait in long lines at the airport, or having to go through stringent security measures just to be rushed to put your bags through more screening, then boarding a full plane with crying babies and uncomfortable turbulence. Imagine not having to take a full day to get to your destination. Imagine no longer having to expel jet fuel. Well, that's what the Dene did in the days of medicine power. They knew how to time-travel. There were no dangerous highways slick with oil after a fresh rain because there was no need for cars, which meant no car crashes, no railroad systems, no need for large tanker vessels to cross the seas, no need for gas stations.

In the times when time travel was implicit among the Dene, entire camps of thousands of people would disappear in a matter of seconds if they knew their camps were about to be bombarded. The only evidence left behind indicating that people were there would be the ash of their smoldering fires. If you still don't believe it just look to the law of physics. Einstein's theory of relativity says that time is measured by our perception. If it is possible through science to bend time, then it's not that farfetched to think that holes in the rock in Dënéndeh and Dinétah could be portals or black holes which have been proven to exist.

The story of Spider Rock, one of Dinétah's most sacred landmarks, tells of a woman who was the first weaver—could it be that the world was once sewn together by one large thread? I can envision an invisible needle and thread sewn through the holes in the rock by Mother Earth just like how Mary Jane Cazon says we are to sew the womb that once held our children. Has that invisible thread become unraveled? If so, how do we sew ourselves back together?

Toward the end of our trip to Dinétah we drove through Sedona, and on the drive down the steep winding hill into the red sandstone valley below, I realized I had already been there in my dreams which I believe is a form of time travel.

Our last leg of the trip to Dinétah included stopping at the Grand Canyon. Words cannot describe the intensity of standing on the edge of the world. It stopped me in my tracks when I got close to the ledge. My daughter describes looking out at the canyon as though looking at a painting—with so many colors and dimensions it was dizzying. We did a bit of research on the statistics of how many tourists visit the Grand Canyon each year, and found that people have suffered cardiac arrest at the edge of the 10,000-foot cliffs from the powerful, overwhelming feeling of just standing near it. Watching the ravens dip in and out of the canyon and coast in the crosswinds I noticed that they too were the same size as the ravens in the north.

Sometimes I have high expectations when I visit certain places in the world that I am told are breathtaking only to be let down by the commercialization of those places, so I did not know what to expect when visiting the Grand Canyon. Now that I have seen it I was not disappointed, I think it's important that everyone has a chance to see it with their own eyes at least once in their lifetime, as it's a very humbling experience.

We had a hotel that was mere feet from the ledge of the canyon in the park, and before falling asleep that night my daughter and I felt a strong spiritual energy—a pull. We knew we were in a very sacred place bordering somewhere between two worlds that was not to be taken for granted. The Indigenous peoples of that area, the Hopi, believe that the canyon is a path to the underworld, and I don't doubt it. Being on the very border of the canyon was similar to the feeling I had after swimming in a cenote in a small village on the Yucatán Peninsula in Mexico a few months earlier. The village was small, with no more than fifty or so inhabitants.

A cenote is a sinkhole with fresh groundwater. Each cenote in Mexico is connected by an underground river system and all are sacred to the Indigenous peoples of Mexico. I was hesitant to go into the cenote. I could feel that there was a very powerful energy force that I did not want to interrupt.

Before going into the cenote, I cautiously stood for a longtime knee-deep in the water on the stairs leading into the deep end, unsure of whether I should go on any further. The water was clear and about six feet deep all around the center of the circular cave that was once used as a well by the villagers. I felt a slight pinch on my knee and looked down to see hundreds of small black fish a bit larger than minnows swarming and nipping at my legs. I took it as my invitation. I slowly lowered myself into the water and swam in and around the limestone ridges, looking at the fractures in the rock. Some parts of the cenote were quite dark, never to see the light of day, and the rocks in the shaded parts of the water began to look like reptiles. I kept swimming even through my fear, with my daughter following behind me like a mother with her baby duckling. We swam around the cenote three times in a clockwise direction. On the last go-around I looked closely at the only petrified tree in the cenote, where a shrine had been made at the entrance of a small hole in the ceiling of the rock. As I was looking into the hole, half expecting to see something, a bat flew out and swooped down above the water to my left. To the Mayans cenotes are portals to the underworld or the afterlife, and bats are symbolic of darkness and the supernatural. That night I felt the energy in my chest just like I did at the Grand Canyon.

That night the wind roared and didn't let up until late the

next morning. It rained the rest of our trip, and I feel that it might have been because we clapped a bit too much to the rain god at Chichén Itzá. While at the Mayan ruins I learned that the Mayans also believe in two brothers who saved the people from giant animals. The story is contained inside what is known as the Popol Vuh, the laws of the Mayans. I found many of the Mayan's cultural practices to be similar to ours back home.

James, the healer, says that Indigenous people are the belt that wraps around Mother Earth, that we've always been connected.

Standing under the watchful eye of Window Rock on that windy day as a tumbleweed rolled by and snagged onto my skirt, healer James told me that the reason why we are in a current climate crisis is because many of our people are no longer doing ceremony. If we were all intentionally praying to Creator for rain when it is needed, we would not be in a drought. This would be the workings of medicine power. Medicine power is referred to as įk'ǫǫ̀ in Tłı̨chǫ Yatıì.

Dene author George Blondin writes that medicine power has become weaker over time because Mother Earth is no longer free from contamination and destruction and is getting worn and old. But medicine power is not something that has been lost forever, says Dr. Redvers, "it's just sleeping right now." Dr. Nicole Redvers has never seen as much interest in Indigenous knowledge—as it pertains to climate change, conservation and biodiversity efforts—as she does now. There are pros and cons to that, but the end point is that there are more people looking, listening and learning from Indigenous people on what needs to be done to help Mother Earth which is a good sign says Redvers. Ultimately, she says, we are seeing a young generation coming up that's much more attuned to and has much more empathy for the global warming situation, and that young generation truly believes what their Elders are saying about medicine power coming back at a time when change is needed.

Dr. Redvers has heard many times from Elders that the new generation of Indigenous people will be the healers. Like what Stacey Sundberg says about the children being reincarnated, the ancestors are coming back to help and guide us all during this time of crisis. "Medicine power hasn't disappeared, it's just sleeping. It has the potential to be woken back up."

"Medicine power hasn't disappeared, it's just sleeping. It has the potential to be woken back up."

One way to wake it up is to begin using it again, to strengthen those muscles that haven't been used in a while.

Elders and the youth need to coalesce to complete the knowledge exchange says Redvers. I wonder why it is still sleeping, and then it dawns on me—we get our medicine power from the land, and because we are getting further and further away from that connection to the land, we are losing our medicine power more and more. Which is why it is so important that we spend more time on the land as much as possible.

There are many forms of medicine power and many different Indigenous spiritual medicine practices across the world. Just like there are different dialects for the same language, there are also different medicine teachings.

George Blondin writes that some people are born with medicine power, while others are gifted it in their lifetime, if they are open to receiving it and others hone it over time. I believe this to be true, and I also believe that it will be bestowed upon a person if they are living a good life but the one thing to remember about medicine power is that if a person has it they will not go around saying that they have it nor will they charge money to help others. There are many people who go around calling themselves healers who are not healed themselves so when seeking out a medicine healer it is important to make sure they have been verified by their community. Above all else, medicine power should never be used for ill intent.

The Dene were once able to shapeshift or transform ourselves from man to animal and see through the eyes of animals. Some were able to melt metal. Some medicine people had the power to break rocks with their words, and others could hide themselves inside of rocks explains Blondin in his book *Yamoria the Lawmaker*.

Some had telekinesis and can move objects with their mind. Some had medicine so strong that when they yelled it can stop a grizzly bear in its tracks. Some people were given the gift of dream visions to help them while on the land. Some are able to read minds, and if you think this is absurd just look at modern-day science where brain-scan technology is able to read a person's thoughts. Laws are already being put in place to ensure that thoughts are safeguarded from theft once the new technology is available more widely.

Some have the power of the insect. Some are gifted with the power of the elements and can create blizzards or high winds. Today this is what geo-engineers are capable of. There is one story in particular as told by Johnny Neyelle in his book *The Man Who Lived with a Giant*. The story tells of how a man named Sahlie had the power to harness the sun. Since *Sah* means "bear," in Sahtúǫt'įne Yatį́ and it was Bear that stole the sun than you can put two and two together and know that these stories are not coincidence, they are truths from a buried history.

If at one point we had the power to call on the elements then it is possible to call on medicine power again to help protect Mother Earth like James suggests.

The one thing we may not be able to do is live forever. There was a medicine man who once tried to make himself young again, says George Blondin in the story "The Man Who Challenged Old Age" in *When the World Was New*. He made himself young for a few days but became old again. He then declared that no one would ever have the power to make themselves live forever, because if he couldn't do it no one could. Nature has a life cycle and so do humans. Eventually everyone will die. The man in the story predicted that humans would one day try to use science to make people live longer, but they would not succeed. There is no sense in being afraid of death. We all will die one day and when we do I believe that we will enter a new world, one that we dream of. All that death is, is a deep sleep. If we choose to come back to this world, we can reincarnate ourselves. Some choose to come back to handle unfinished business or come with a specific purpose in life.

Steven Nitah says there is a place where people can go to regenerate themselves or be reborn. "It's a little lodge. You go in there and you come back as a baby." The only thing that can't regenerate is the teeth. This is why Elders say that babies born with teeth are those that regenerated themselves explains Steven.

Fibbie Tatti recalls going with her grandfather every evening to one of the prophets' houses in her community. The house would be so full sometimes that it was standing room only, so she would have to sit perched on a woodpile.

Before prophet Ayah there was another prophet who saw the future as well. People came to him from different communities to hear what he had to say. The prophet shared many stories with Chief Fred's grandfather, who was born in Sathú—one of which was a warning that European people were coming and that their way of life was different than the Dene way of life. "They are going to come to you, come to your land and take your land. They will work on the land and destroy it. The land will be destroyed, the water will be destroyed, everything will be destroyed," he says, and "it will end with fire."

Fred informs me that there was a promise made between Creator and Indigenous people a long time ago. It was the first promise ever made between man and Creator. The promise was that we would respect Mother Earth and not destroy anything. To this day, Indigenous people all over the world are defenders of the land, says Fred. Even if the leaders of the Indigenous nations want to make great profit from industry, there are still people within those nations that will defend the land. An example of this is when hereditary leaders and elected leaders within the same nation clash. But as part of the original instructions Indigenous people were given, "it's built inside us. It's within us to protect and care for Mother Earth," says Fred. What the prophet shared long ago "was a sacred omen," Fred tells me. The land will start burning because the agreement has been broken. We didn't take care of Mother Earth, and Mother Earth cannot be replaced. "We'll have no food, no firewood… nothing to grow. We'll have no wildlife. Everything on earth… will be basically gone. People will be chasing each other. There will be chaos for water. People will fight for water at the very end. Even killing. War. Humans will destroy humans." This chaos resulting from human-made climate change that Fred describes is already happening. "It's just beginning. We see the landslides, we see the floods, we see the tornadoes. The land up here in Dënéndeh is melting away." The permafrost is melting. When Fred used to travel in the barren lands, there was still two feet of solid ice under the ground. "You could turn the moss over and put your fish in there for a few days because it's all ice. Today you could dig down in the same place four or five feet, there's no ice. The ice is gone."

On a canoe trip in the barren lands, Fred and a group of students, he was helping to guide, came across a small lake that used to be about half a mile long by half a mile wide, but the water receded so much that the lake was reduced to a puddle in the middle of the nearly dried up lake bed. "There was thousands and thousands of fish in this water. The water was disappearing." The young Indigenous students had good hearts, he tells me. They were running back and forth trying to put fish in the water at the next lake over. "They probably did that about twenty or thirty times." They became so tired that they couldn't do it anymore but there was still lots of fish. "You could see within a week or so there would be no water. There'd be just fish. They had to give up, knowing that the fish would eventually die without water and it was heartbreaking for them because they were helpless says Fred. Water is disappearing underground and into the atmosphere, says Chief Fred, the air is dry now—there is no more moisture because fuel has replaced fresh air. "There's a lot of politicians in the world who live in a concrete world, they don't see what's happening outside of that." Climate change is real, says Fred. "We had a warning about this a long, long time ago, where we will come to this part of the history where the land will burn."

We have to adapt to climate change and that's going to cost a lot of money, says Fred. New maps will have to be made. Maps without water.

Cataclysmic events like the ones that Fred speaks of are told numerous times in the *Book of Dene,* and so are stories of giants. Joni Michelle Tsatchia was always curious about giants. Joni is Tłı̨chǫ. She grew up in Wekweètì, a small Tłı̨chǫ fly-in community. Wekweètì is very isolated. The only way people can get in and out is on the ice road in winter or by plane. Wekweètì was Joni's playground growing up. She loves being out on the land. She calls the land her first home. Wekweètì is a place that she would like to retire, but for now she goes back to visit as often as she can. All her childhood friends still live there.

Joni was very close with her grandmother, Marguerite Tsatchia. One day Joni was sitting outside on a blanket of spruce bough beside her grandmother, and there were ants near her. She didn't want them to crawl on her so she was trying to kill them. Her grandmother saw her doing that and said, "You know Joni… that ant is really small and you are making it suffer, it's not really nice." Then her grandmother proceeded to tell her a story that she had heard from her grandmother. "Every animal in this world at one time was a giant, and when the world is going to end, in the end of days, these animals will wake up." Her grandmother called them "the giant ones," and said, "if you are not being very nice to the ants, the giant will remember you and will look for you." Joni then could not get the thought out of her mind that every insect, every bird, every land- and water-dwelling animal was once a giant and she never killed another ant again if she could help it.

What her grandmother taught her that day has never left her. Joni now tells the story of giants to her own children. "When we used to live in the log house there was a lot of spiders," she says. Her grandmother would put the spider on a piece of paper and let it free outside. She would never kill animals unless it was for food. "Even though you think this little insect can't hurt you, if there's a giant for every animal it makes you respect them." Joni's grandmother said the giants were called godę̀ęcho in Tłı̨chǫ Yatıì.

On another occasion, when Joni was driving on the ice road with her grandmother, she looked out at the rocky landscape covered in trees. "There was one hill in particular that got bigger and bigger as you passed by it." Joni remembers pointing to it and saying, "Grandma, look at that big hill." Joni's grandmother replied, "Remember I told you about the giants… well, that's one giant that's going to wake up."

In a story called "The Baby Raised by Godene," George Blondin writes about a giant who was said to be a huge apelike creature. The story starts out with a young couple who had a baby, but the mother died in childbirth and the man had to raise his son alone. One day he had to leave his son to go out and hunt, and when he returned his son was gone. He saw Godene's giant footprints and followed them. There were wood shavings along the way, evidence that Godene was carving snowshoes.

The father caught up to his son after three days. He found him sitting alone shooting ptarmigan. He had grown from a baby into a boy in that short amount of time. The boy's father approached their camp carefully when the giant had gone ahead of the boy to make a new camp for them. When the father told the boy to come with him, the boy refused. He was tricked into believing that Godene was his father, but after his real father pleaded with his son the boy understood that he had been kidnapped by the giant. The boy and his father then made a plan to kill Godene. They clubbed him over the head when he was least expecting it. The boy and his father returned to their own camp. The people in the community revered the boy because he was very helpful and they knew he possessed strong medicine power because of what had happened to him.

I have been told that our people walked with giants in the time of the ice age and I believe it. Stories of giants can be also found in the Dëne Sųłıné chapter of the *Book of Dene.* An Arctic giant named He Whose Head Sweeps the Sky and another named Bettsinuli once got into a fight, and although Bettsinuli was the stronger of the two giants, He Whose Head Sweeps the Sky had help from his friend, a man, who cut Bettsinuli's leg with an axe made out of a giant beaver tooth. Bettsinuli fell backward into the water and turned into a mountain, which eventually became a place to where the caribou migrated. There is also a story in the *Book of Dene* where the giants were inside a mountain and a man turned himself into an ermine. The ermine threw a piece of burning wood into the mountain and the entire mountain caught on fire and wiped out all the giants. From then on, it is said, there were no giants left on earth—but some say they only went into hiding and will one day return, like Joni's grandmother cautioned.

There are giants and then there is also the belief in an entity called the Nàhgą in Tłı̨chǫ Yatıı̀, which translates to "bushman." The giant and the bushman, otherwise known as sasquatch in some parts of Canada, are one and the same in some Dene communities. It is said that if children are out too late at night, they might get taken. My own grandma warned me when visiting Nı̨hshìì not to roam too far and to stay away from the tree line or the Nàhgą might capture me.

I have a notion to believe that the bushman can shapeshift not only into animals but into trees and rock. I have been told that sometimes we are able to see glimpses into another dimension where sasquatch live when we are at pivotal turning points in our lives. The day my son left home was one of those times.

My only son was leaving the nest. He wanted to move back north, and even though it was bittersweet for me to say goodbye to him, I felt he was ready to be on his own. My daughter and I drove him to the airport at an ungodly hour when it seemed that no one else in the world was awake. We said our tearful goodbyes—it had been the three of us for so long and now we were down to two. My daughter and I headed back down the road toward home on the dark and lonely highway in silence. We were in an area right beside one of the First Nation reserves, and there are many stories of the sasquatch in that neck of the woods.

I was driving along at a bit of a slower pace, because I don't like driving in the dark as I'm afraid I might hit an animal, and that's when we saw something fall into the road right in front of our vehicle. It looked like a large bushy tree about two and a half meters tall, but it was a transparent grey colour and looked almost like static on a television screen. I tried to swerve out of the way so as not to hit it, but I didn't have time and that's when we saw another tree figure appear behind it from out of thin air. It all happened so quickly as we watched the fallen tree being lifted up off the road as though it was being leant a helping hand. I looked at my daughter in shock.

"Did you see that?" She just nodded wide eyed. I wasn't about to stop and get out to see what it was so I just kept driving and looked back in my rearview every so often.

Maybe that is how the sasquatch hides so well in the temperate rainforest. I believe that maybe my daughter and I experienced this sighting because it was a new chapter of our lives, where we would have to help one another and pick each other up without my son there with us. Or could it also be that deforestation and other environmental impacts are destroying forests and sightings of the bushman are increasing? These two truths could very well exist in tandem. Every year there are thousands of bushman sightings across North America and all over the world and they are only increasing.

In Dënéndeh it is said that when the fog rolls across the open water or open fields, that is when the bushman is traveling because he cannot be seen through the fog. That is why it is also said that it is not a good time for hunters to travel when it's foggy outside.

There are medicine people who can create fog or mist. In *When the World Was New,* George Blondin tells the story of how a medicine person heated up the skin from the head of an owl until it became soft, then he put it on his own head and sang until a fog rose. He created fog so that the people who needed to cross the land to safety to get away from their enemies were able to quietly slip away in the mist by holding onto a rope and walking in a straight line together.

I have been cautioned not to talk too much about the *Nàhgą,* but I feel better knowing that Elder Jonas spoke about the bushman without fear—in fact, he found it amusing to retell the story about how a bushman was once killed. Even Berna was surprised by what Jonas spoke of while she was translating the conversation in his living room on that dark winter's evening when we went to visit him.

There was an old man in Wekweètì who heard voices—strange unfamiliar voices Jonas says. Out of nowhere, three bushmen started chasing the old man. The old man managed to shoot one of the bushmen dead but only shot the tip of the ears of the other two, so they survived and got away. The old man cut the dead bushman into pieces. In those days people wanted the spirit of the bushman to go further away from them, so they were doing all kinds of things to chase the bushman away—including using medicine power. The old man told his son to stick the pieces of the bushman's body on the top of a sharp stick and stake it into the ice of the lake. When the son did as he was told, the seagulls flew down by the thousands and feasted on the bushman's carcass.

The old man was a tough man, very strong, says Jonas, and that was because of the medicine he had. He traveled alone using his medicine power and he survived through almost everything but the old man died peacefully with the Elders at his bedside shortly after he had killed the bushman. Before he died he would sleep with a gun under his pillow inside his tent and he would cover himself in ash from the embers of the fire for protection.

My great-uncle Alfred can understand some English, and so when I ask Berna to ask him if he knows what the future holds for humanity and Mother Earth, he answers me quickly in English without her help. The damage is done, he says. But nobody knows when it is going to be the end of the world. For all we know there might be something that will fall to the earth and destroy the whole earth or maybe artificial intelligence will be our downfall I add but Alfred's says, only Creator knows. People that talk about the end of the world are predicting their own death, he says. It's hard to predict what is going to happen. It's hard to say how things will be in the future. He points to me and says that I will probably still be walking this earth when I am an Elder. He points to his great-grandson who is running back and forth from the kitchen table where we are sitting to the couch in the living room, where his cartoons are on. Every so often he runs and crouches under the kitchen table and playfully tries to hit the keys on my computer as Alfred looks on and laughs at the energy of his youth. Maybe he sees himself in the busy toddler's eyes.

He too will grow to be old, says Alfred. "He's like a little weasel," he teases with a squint of happiness in his eyes. Toward the end of our conversation Alfred's voice becomes strained and he can no longer talk. He was close to ninety years old then and one of the last of the drummers in the community to sing like the old ones.

Alfred has since passed on into the spirit world, but his memory will live on through his children, grandchildren and great grandchildren. People tell Randy Baillargeon all the time that he sounds like his grandfather, Alfred, when he sings. "He gave me a gift," he tells me. That style of singing has almost been lost, I say, and Randy laughs and says, "Not anymore." Randy is thankful for the songs that he learned from his grandfather and other Elders in the community. He will pass the old songs onto his son Ahfrè as well one day.

Randy says "One of these days there will be no more oil, no more gas, no more gold, no more diamonds, so what are they going to do? The only way is to live off the land." That's why Ayah and many Elders who have since passed on always said "don't forget who you are, teach your children to live off the land"—and that is why Randy is teaching his son to live off the land.

When my friend Jennie McPherson thinks about climate change, she thinks about Ayah. Ayah, she tells me, had a lot of prophecies about climate change—specifically how we would see a lot more earthquakes, famine, hurricanes, floods, drought in the future. His prediction, Jennie says, was that "if we don't love and respect each other, the land and ourselves, then that's the path we're headed down. You need to have a good healthy relationship with yourself, so that you can have that with others and also with the land, because says Jennie "we believe that the land and the water and everything is alive, so it's about having a good relationship with it."

Ayah's prophecy is a warning. "People often think prophecies are a done deal, but they are not." It's a glimpse of what can happen if we stay on this path, says Jennie. If we can heed Ayah's warning and really change how we see our relationship with the world then maybe we'll have a chance. "Whatever we do will impact us eventually, everything comes back around, it's always a cycle, nothing is ever disconnected, everything is interconnected." We must first understand that she says. Part of the solution is a matter of "figuring out who we really are and what our purpose is for being here."

It was starting to get cold on the boardwalk as the sun dropped below the horizon but before parting ways with Dene lawyer Jennifer Duncan she tells me that she grieves for Mother Earth. What can we do for our spirits in the face of this crisis? I ask her. Her answer is that she believes in her heart that the younger generation will have the answers and solutions—they are the ones who are going to find the key. She stops walking and can no longer hold back her tears. She apologizes for getting emotional. I try to think of something to say that will comfort her: "There's going to be some sort of new technology that's going to fix it hopefully. We will be able to adapt. We've adapted before. We are very resilient..." I ramble. That solution might just come in the form of Indigenous futurism. "We might even discover a way to bring back medicine power that allows us to tap into quantum physics," I say with hope. "I really think that quantum entanglement is what our Dene stories are."

Jennifer agrees. "It's more like a concentration of medicine power, it exists." Just because we can't see something doesn't mean it's not there. "We know it's here," she says. She laughs because it sounds like we are a couple of characters in a *Star Wars* movie. "Let the force be with us." I say and we both laugh through our grief for Mother Earth. "I still very much believe in medicine power," Jennifer says. It is through medicine power that we can raise the frequency of climate consciousness.

Unless we collectively shift our consciousness and enter a new narrative where our senses are awakened, we will not escape the impending doom of climate change. Maybe that is what medicine power is—a collective reawakening of our spirits entering into a new reality. To not be afraid of the death of the world as we know it because Mother Earth is reincarnating herself.

As I look out of Celine Marlowe's living room window and see the large water truck nearly running over the government worker as it's being towed backwards by a tractor, Celine explains that it broke down during Covid and ever since then it can only drive in reverse.

"All my life I've lived here and I see all the things that are going on, and the only thing I can say is be prepared for what's coming. Always be prepared. No matter what. Make sure you always have your belongings packed ready to go." Her husband always told her to prepare ahead. Tomorrow she wants to go to her camp, so she's getting ready today. "Always be sure to prepare yourself for what's coming."

We can never know what tomorrow brings us—so pray, be resourceful, be prepared and walk softly on Mother Earth.

Acknowledgements

The connection to the community that I have worked so hard to build is one of trust, and without relationality this book wouldn't have been possible. The storytellers trusted me with some of their most memorable experiences, and I shared in their moments of joy, fear and tears. Through these connections with family and friends I was also able to receive the answers to some of life's greatest mysteries—answers that I didn't even know I was searching for until I embarked on the making of this book. If there's one thing I've noticed on this journey of writing *Mother Earth Is Our Elder*, it is the generosity that my Dene family has shown. They graciously made time to meet with me when I reached out and asked them to share their stories, knowledge, teachings and family histories. They have not only opened up their hearts to me, they opened their hearts to the world for the betterment of us all. I am so grateful that I have been entrusted to carry and pass their messages on. I am so filled with love and gratitude for the gift of being able to share with the world the incredibly powerful lessons that are in this book.

This knowledge exchange is the highest honor, and I feel very humbled.

Putting this book together felt like Christmas morning as a child.

There are so many more Dene voices in the north that have their own stories to tell, and I hope this book sparks further discussion where we can collectively come together as one family and learn from one another to share our knowledge for the greater good.

Mahsi cho to the librarians at the West Vancouver Memorial Library for giving me the idea to write this book in the first place.

Mahsi cho to the healers from Dinétah who gave me sage advice. I will come visit you again soon.

Mahsi cho, Agnes Pascal, for connecting me with the women's sewing circle in Iñuuvik.

Mahsi cho to my family and friends for walking the dog on the days when I was so immersed in writing that I didn't even have time to look up from my computer, and thank you for holding down the fort while I was traveling the north.

Mahsi cho Jennie and James for your wonderful hospitality on the shores of Sahtú.

Mahsi cho to my daughter for helping to compile the stories from the *Book of Dene* into a categorized table.

Mahsi cho to my son for giving me a glimpse what it is like working on the front lines of wildfire. Firefighters truly are heroes, and I am proud that I raised one.

Mahsi cho to Marie Coe for keeping Aliviya and I company on our road trip across Dinétah.

Mahsi cho Maro Sundberg for translating words into Wıı̀lıı̀deh Yatıı̀ and mahsi cho Berna Martin for interpreting during visits with the Elders.

Mahsi Cho to Dr. Renae Watchman for taking the time to review the manuscript with a focus on references to the Diné.

Mahsi cho Gabriella Page-Fort for believing that I am the right person to write this book.

Mahsi cho Stephanie Sinclair for championing this work and Kelly Joseph for your kind words of encouragement along the way and the late nights of going back and forth with me on edits.

Mahsi cho to my agent Cody Caetano and former agent Rachel Letofsky for helping to draft the initial proposal and send it out to the publishers.

Most importantly, mahsi cho to all the storytellers who shared their valuable insight. In no particular order they are Stacey Sundberg, Melaw Nakehk'o Antoine, Tanya Lantz, Darrell Beaulieu, Don Balsillie, Wanda Pascal, Joni Michelle Tsatchia, Steven Nitah, Lee Mandeville, George Mandeville, David Sangris, Trenton Coe, Dr. Nicole Redvers, Cassandra Blondin Burt, Lawrence Casaway, Cindy Allen, Chief Fred Sangris, Celine Marlowe, Daniel T'seleie, Mary Jane and Gilbert Cazon, Jennifer Duncan, Jennie and James McPherson, Tommy Lafferty, Paul Mackenzie, Maurice Zoe, Randy Baillargeon, Alfred Baillargeon, Jonas Noel, Gladys Norwegian, Alec Rabesca, Letitia Pokiak, John B. Zoe and Fibbie Tatti.

I turned the first draft of this book into the publishers on the evening of an earthquake that woke me up out of my sleep on the west coast. If that wasn't a sign from Creator to prompt me into action to get this book done, I don't know what is.

Mahsi cho Creator for all the signs along the way and for guiding me on this lifelong journey of enlightenment.